Apocalyptic Visions in the Anthropocene and the Rise of Climate Fiction

Apocalyptic Visions in the Anthropocene and the Rise of Climate Fiction

Edited by

Kübra Baysal

Cambridge
Scholars
Publishing

Apocalyptic Visions in the Anthropocene and the Rise of Climate Fiction

Edited by Kübra Baysal

This book first published 2021

Cambridge Scholars Publishing

Lady Stephenson Library, Newcastle upon Tyne, NE6 2PA, UK

British Library Cataloguing in Publication Data
A catalogue record for this book is available from the British Library

Copyright © 2021 by Kübra Baysal and contributors

All rights for this book reserved. No part of this book may be reproduced, stored in a retrieval system, or transmitted, in any form or by any means, electronic, mechanical, photocopying, recording or otherwise, without the prior permission of the copyright owner.

ISBN (10): 1-5275-7305-2
ISBN (13): 978-1-5275-7305-5

TABLE OF CONTENTS

Introduction vii
Kübra Baysal

Chapter One 1
A Tragic Glimpse of Thoreau's Prophetic Vision in *Walden*
Onur Ekler

Chapter Two 13
Pre-Histories of Cli-Fi: The Agential Reality of Richard Jefferies's
After London
Adrian Tait

Chapter Three 29
An Ecocritical Analysis of the Natural Life in Ursula K. Le Guin's
The Dispossessed (2002)
Seher Özsert

Chapter Four 46
The Apocalypse of Free Will in *The Dispossessed*
Pınar Süt Güngör

Chapter Five 61
The Post-Apocalyptic World Order in *The Tin Can People*
by Edward Bond
Elvan Karaman

Chapter Six 78
Transforming Bodies in Maggie Gee's *The Ice People* and HBO Max's
Raised by Wolves
Niğmet Çetiner

Chapter Seven 93
Society Islands and the Anthropocene in David Mitchell's *Cloud Atlas*
Emily Arvay

Chapter Eight 111
The 'Dying' Earth in the Postnatural Age of the Anthropocene:
An Ecological Reading of Amitav Ghosh's "Cli-fi" Sequel
Risha Baruah

Chapter Nine 129
The Ecological Conscience of Ian McEwan's Fiction
Anastasia Logotheti

Chapter Ten 145
Care More, Destroy Less: The Environmentalist Ethic of Care
in Liz Jensen's *The Rapture*
Işıl Şahin Gülter

Chapter Eleven 164
Monstrous Agency and Apocalyptic Visions in Paulo Bacigalupi's Climate
Change Trilogy: *Ship Breaker*, *Drowned Cities*, and *Tool of War*
Sukanya B. Senapati

Chapter Twelve 182
Revelations of the Millennium: Posthuman Beings and/ in Postapocalyptic
Worlds
Andrew Erickson

Chapter Thirteen 201
The Post-Apocalyptic Aesthetics of Emily St. John Mandel's
Station Eleven
Murat Kabak

Chapter Fourteen 213
Failing Ecosystems, Contagion, and Catastrophe in Zadie Smith's
'Narratives of Entropy': *The Canker* and *Elegy for a Country's Seasons*
Manasvini Rai

Chapter Fifteen 233
Have We Lost Our Senses?
Roma Madan-Soni

Contributors 260

INTRODUCTION

KÜBRA BAYSAL

The twenty-first century has seen the manifestation of the Anthropocene, which entails negative environmental phenomena such as climate change, gradual loss of biodiversity, disruption of the ecological balance, and deterioration of the Earth system[1] in line with the rise of technology, overpopulation, and the launch of the space age for the exploration of life on other planets, all of which constitute the global environmental agenda. Advanced technology has empowered humankind while destroying the nonhuman *environs*, which, since the beginning of the century, has left scientists, environmental humanities scholars, and writers concerned and has led to research and numerous papers on the subject. Simultaneously, the popularity of 'end of the world' scenarios, reminiscent of the *fin de siècle* in the nineteenth century, has significantly expanded in works of both science and literature since researchers and writers increasingly began to voice their concerns regarding the jeopardized future of humankind along with the destruction of the nonhuman world, and its devastation through common catastrophes, illnesses, human strife, and savagery. In this light, prominent climate change fiction novelists who have contributed to the canon, i.e., Margaret Atwood, Ursula K. le Guin, Octavia E. Butler, Doris Lessing, Jeanette Winterson, Kim Stanley Robinson, Liz Jensen, Maggie Gee, Barbara Kingsolver, Marcel Theroux, Ian McEwan, Paolo Bacigalupi, T.C. Boyle, Sarah Hall, David Mitchell, Amitav Ghosh, Saci Lloyd, Nathaniel Rich, and Adam Nevill, among many others, speculate on the great fear of humankind about their end as part of the *en-masse* annihilation of the nonhuman world.

---

[1] Earth system is a term referring to the "physical, chemical, and biological processes that transport and transform materials and energy and thus provide the conditions necessary for life on the planet" (see Will Steffen et al. *Global Change and the Earth System: A Planet under Pressure* (Berlin: Springer-Verlag, 2004), 8). The term reflects the phenomenon of life in the world and is more generic than climate change itself for it includes all kinds of natural and human-induced transformations on Earth.

Giving rise to the fears of humanity, the Anthropocene has become an undeniable reality against whose impact it is imperative to take measures by all humanity in a collaborative effort. One of the most visible effects of the Anthropocene in the biosphere detected from the late eighteenth century to the 2010s is the increased release of CO_2 (carbon dioxide) by "a factor of 5.4" (1860-2014) combined with nitrous oxide produced by supersonic transports and odourless CFCs.[2] In other words, damage caused by humans to the biosphere is beyond measure, and it has been affecting and transforming the Earth system and its biodiversity completely. While carbon emissions rise, humanity arrives at threshold points from which there is no possibility of returning to the "homely comforts of the Holocene."[3] It is predicted with high probability that the degradation in the biosphere, or "Earth's living tissue,"[4] through high carbon emissions will trigger a rise in temperatures of 2 °C in the near future, which will increase the melting of the few remaining ice sheets in Greenland and the Arctic Sea and rising sea levels even more dramatically.[5] The deterioration in the biosphere has tremendous effects not only on the Earth system and the nonhuman environment in the form of drastic transformation and extinction but also on human beings through new strains of viruses and pandemics such as Covid-19 and social, economic, and political turmoil, because humankind has always been intra-active with its surroundings and is now on the verge of the sixth extinction as "a casualty of [its] history."[6] In this regard, humanity in the twenty-first century faces the long-term impact of the centuries-long anthropogenic transformations on Earth and feels the urgency to take immediate action much deeper than ever.

Coined by North American activist and journalist Dan Bloom in 2007, cli-fi is the term for works that focus on anthropogenic environmental problems and their diverse effects on human beings. It can be explained with the keywords "contemporary, controversial, transmedial, transnational, didactic, generic, [and] political".[7] Climate change fiction, which as an

---

[2] Christian Schwägerl, *The Anthropocene: The Human Era and How It Shapes Our Planet* (London: Synergetic Press, 2014), 5, 39.
[3] Brad Evans and Julian Reid, *Resilient Life: The Art of Living Dangerously* (Cambridge: Polity Press, 2014), 3.
[4] Christophe Bonneuil and Jean-Baptiste Fressoz, *The Shock of the Anthropocene: The Earth, History and Us* (London: Verso, 2016), 6.
[5] Brad Evans and Julian Reid, 143.
[6] Eileen Crist, "On the Poverty of Our Nomenclature" *Environmental Humanities*, 3 (2013): 137.
[7] Julia Leyda, "The Cultural Affordances of Cli-fi," from "The Dystopian Impulse of Contemporary Cli-Fi: The Dystopian Impulse of Contemporary Cli-Fi: Lessons

umbrella term encompasses Anthropocene fiction, is interdisciplinary and international as climate change is obviously not restricted to one field of study or to the borders of one country and deals with the environmental problems that have been growing in urgency since the twentieth century. As Johns-Putra asserts, cli-fi is a significant "literary and publishing phenomenon" and "a category of contemporary literature", more of a topic than a genre as it can be observed in numerous genres such as "science fiction, dystopia (themselves two genres given to much cross-fertilization), fantasy, thriller, even romance, as well as fiction that is not easily identifiable with a given genre, for example, the social or psychological character studies favoured by mainstream authors such as Maggie Gee, Barbara Kingsolver, and Ian McEwan."[8] For this reason, since it is more popular in the novel genre, it moves beyond the conventional novel form as it lays emphasis on the impact of environmental changes on humans rather than on generic qualities such as plot, unity, or character development, and "exceeds what the form of the bourgeois novel can express."[9] Its focus is on depicting the factual environmental conditions in the fictive world of the novel most effectively, thereby conveying grave messages for individuals and political bodies to take action against the environmental transformations before it is too late to act. In this respect, the primary function of cli-fi is "[c]reating a connection between the reader and characters immersed in disastrous global warming, [through which] readers could immediately experience climate change as a threat to their centers of felt value."[10] The effect of environmental reality appears to be more significant when those facts ornament the general outline of the novels in which human beings tend to associate themselves with the protagonists and feel for them as they are struggling with difficulties, namely natural disasters, or worrisome environmental problems. At this point, cli-fi is the recent form of the novel genre that emerged out of the necessity to discuss environmental issues from a non-anthropocentric and posthuman perspective:

---

and Questions from a Joint Workshop of the IASS and the JFKI (FU Berlin)" (Institute for Advanced Sustainability Studies (IASS), 2016): 12.

[8] Adeline Johns-Putra, "Climate change in literature and literary studies: From cli-fi, climate change theater and ecopoetry to ecocriticism and climate change criticism" *WIREs Climate Change*, 2016.

[9] Mckenzie Wark, "On the Obsolescence of the Bourgeois Novel in the Anthropocene" (*Verso* "Authors' Blog", 2017).

[10] Adam Trexler, *Anthropocene Fictions: Novel in a Time of Climate Change*. (London: University of Virginia Press, 2015), 76.

> Many of the traditional features of the novel, which are anthropocentric and conflict with ecological integrity, must be reinvented. To capture the particularities of place and lead to ecological enchantment, they must be creative and employ poetic nature diction [...] We find then that climate change gives rise to a new form of novel, which steadily gives rise to a new way of conceptualising the issue.[11]

In this sense, works of climate fiction which reflect the massive impact of climate change on humankind and the nonhuman have become the key narratives of our century. Pointing out the major part that humans play in the Anthropocene, in other words, the human age, cli-fi ironically reminds the reader of the inconsequential and helpless state of humans in the face of natural disasters and devastation. In Tuhus-Dubrow's words,

> climate change is unprecedented and extraordinary, forcing us to rethink our place in the world. At the same time, in looking at its causes and its repercussions, we find old themes. There have always been disasters; there has always been loss; there has always been change. The novels, as all novels must, both grapple with the particulars of their setting and use these particulars to illuminate enduring truths of the human condition.[12]

As a popular literary category of our time which has obviously descended from the apocalyptic fiction and science fiction of the past centuries, cli-fi tackles the issue of human accountability for the destruction of the Earth by making the reader question themselves in terms of their place in life whilst reading about the associable and realistic struggle for survival of the protagonists in the inhospitable conditions of their world(s). For this reason, climate narratives point out "the historical tension between the existence of catastrophic global warming and the failed obligation to act" and provide humankind with "a medium to explain, predict, implore and lament."[13]

In this spirit, through climate fiction, authors, playwrights, poets, and artists elaborate on the urgent environmental issues for humankind from their individual perspectives in the said eras, which is the focus of this volume. Picturing the emergent climate change and the natural as well as social trauma it entails, cli-fi gives voice to the pressing concerns of humanity. In this vein, with the aim of raising awareness while harbouring

---

[11] Sophia David, "Eco-Fiction: Bringing Climate Change into the Imagination" (PhD Dissertation. University of Exeter, 2016), 13-14.
[12] Rebecca Tuhus-Dubrow, "Cli-Fi: Birth of a Genre" *Dissent* (2013).
[13] Adam Trexler, 9.

hope through associable stories of the protagonists, these works depict dystopian and pre-/postapocalyptic worlds of the past, present, or future stricken by a myriad of climate change calamities. To this end, it is highly functional since it encourages the reader to face and deal with the problems of climate change in their immediate environment and inspires them to make a change in their lives through grim depictions of the undesirable conditions in the worlds portrayed. Offering diverse perspectives with each work, cli-fi provides the reader with the possibility of various emotions "from dystopian despair to glimmers of hope, from an awareness of climate change impacts on generations to come to vivid reminders of how we are destroying the many other species that share our planet."[14] Therefore, it has apparently proven its capability of luring the reader into witnessing the dramatic paradigm shifts in the anthropogenic world of our time while urging them to think and act. Empirical data shows that climate fiction exerts a more permanent impact on the reader than statistics and "scientific facts of drought, sea level rise, and species extinction" because they see the stories depicted in those works as "cautionary tales, not prophecies […], as warnings about possible futures."[15] For this reason, climate fiction holds a key position as a literary phenomenon to awaken the reader to the unrepresented or underrepresented environmental realities of the twenty-first century through the tangible situations and associable adventures experienced by the characters.

Composed of fifteen chapters, this edited volume discusses the rise and importance of climate change fiction as part of popular literature, media, and art through a survey of the forerunners and recent representatives of the genre from the nineteenth century to the twenty-first century. From Henry David Thoreau and Richard Jefferies to Amitav Ghosh and Zadie Smith, each chapter in the volume seeks to dwell on the vital role the climate change phenomenon and climate change fiction itself play in contemporary literature and art as the reflection of life. The contributors, who analyse cli-fi, media, and art works from an environmental humanist perspective, aim to discuss and represent the varied layers of the genre itself. As climate fiction bears the responsibility of conveying to humankind their intricate role in the Anthropocene and encouraging them to make a positive change in the world, this volume attempts to contribute to the field of literature, social sciences, and environmental humanities among countless others. Addressing individuals and political bodies alike and reminding them of the

---

[14] Adeline Johns-Putra, "Cli-fi: The seven most crucial climate change novels" *Quartz*, 2019.
[15] Matthew Schneider-Mayerson, "The Influence of Climate Fiction: An Empirical Survey of Readers" *Environmental Humanities,* Vol. 10, No. 2 (2018): 485-486.

interconnection of human and nonhuman in the face of ongoing environmental transformations through an array of climate fiction analyses, this collection strives to call attention to the realities of our time and reaffirm the value of literature and artworks for the critical messages they deliver to humanity.

References

Bonneuil, Christophe and Fressoz, Jean-Baptiste. 2016. *The Shock of the Anthropocene: The Earth, History and Us*. London: Verso.
Crist, Eileen. "On the Poverty of Our Nomenclature" *Environmental Humanities*, 3 (2013): 129-147.
David, Sophia. 2016. "Eco-Fiction: Bringing Climate Change into the Imagination" PhD Dissertation. University of Exeter. https://ore.exeter.ac.uk/repository/bitstream/handle/10871/24331/DaviдS.pdf?sequence=1
Evans, Brad and Reid, Julian. 2014. *Resilient Life: The Art of Living Dangerously*. Cambridge: Polity Press.
Johns-Putra, Adeline. 2019. "Cli-fi: The seven most crucial climate change novels" *Quartz*. https://qz.com/1770404/the-seven-most-crucial-climate-change-novels/.
—. "Climate change in literature and literary studies: From cli-fi, climate change theater and ecopoetry to ecocriticism and climate change criticism" *WIREs Climate Change*, 2016. doi: 10.1002/wcc.385
Leyda, Julia. "The Cultural Affordances of Cli-fi," from "The Dystopian Impulse of Contemporary Cli-Fi: The Dystopian Impulse of Contemporary Cli-Fi: Lessons and Questions from a Joint Workshop of the IASS and the JFKI (FU Berlin)" *Institute for Advanced Sustainability Studies (IASS)* (2016): 1-28.
Schneider-Mayerson, Matthew. "The Influence of Climate Fiction: An Empirical Survey of Readers" *Environmental Humanities,* Vol. 10, No. 2 (2018): 473-500.
Schwägerl, Christian. 2014. *The Anthropocene: The Human Era and How It Shapes Our Planet*. London: Synergetic Press.
Steffen, Will, et al. 2004. *Global Change and the Earth System: A Planet under Pressure*. Berlin: Springer-Verlag.
Trexler, Adam. 2015. *Anthropocene Fictions: Novel in a Time of Climate Change*. London: University of Virginia Press.
Tuhus-Dubrow, Rebecca. 2013. "Cli-Fi: Birth of a Genre" *Dissent*.

Wark, Mckenzie. 2017. "On the Obsolescence of the Bourgeois Novel in the Anthropocene" *Verso* "Authors' Blog." https://www.versobooks.com/blogs/3356-on-the-obsolescence-of-the-bourgeois-novel-in-the-anthropocene?fb_comment_id=1285539134907990_1286166938178543 [20.04.2018].

CHAPTER ONE

A TRAGIC GLIMPSE OF THOREAU'S PROPHETIC VISION IN *WALDEN*

ONUR EKLER

> Amid the moving pageant, I was smitten
> Abruptly, with the view
> Of a Blind Beggar, […],
> Stood propped against a wall, upon his chest
> Wearing a written paper, to explain
> The story of the man, and who he was.
> […][1]

Introduction

Since the early stages of human history, Thoreau has been one of the few who have had a prophetic vision of the impending threats of industrialization. Propped against modern human history with his *Walden*, Thoreau is one of the key figures to light the wick of awareness among deluded people by the spell of mechanical power. In this context, Thoreau may resemble Wordsworth's blind beggar in his *Prelude* who removes the blinding veils from the speaker's eyes with an apocalyptic note upon his chest. That striking note makes the speaker experience a sudden rupture from the illusory world of the funfair. Unlike Thoreau's resemblance to Wordsworth's blind beggar in this respect, the power-drunk people in the newly industrial world seemed to bear no affinity with the speaker in Wordsworth's *Prelude* who unveils the illusion, albeit instantaneously:

> This label seemed of the utmost we can know,
> Both of ourselves and of the universe,

---

[1] William Wordsworth, *The Prelude: Or, Growth of a Poet's Mind; an Autobiographical Poem* (D.C. Heath, 1888), vii 637-42.

> And on the shape of that unmoving man,
> His steadfast face and sightless eyes, I gazed,
> As if admonished from another world.[2]

    Too much trust in scientific rationality, primarily believed to boost human progress and welfare to unprecedented levels and also to make the world a more sustainable habitat, has long blinded people. However, humankind's insatiable desire for dominance and the urge to control everything has completely destroyed such utopic ideals of positivistic thinking at its birth. Each revolutionary intervention of science in the evolutionary functioning of the planet to better the conditions, results in catastrophic events. The devastating effects of science on the planet through sudden and immediate actions have even led to the coinage of a new term for the geological shift of our own creation: Anthropocene or it is better to call it "anthropo[sin]" since the universe is dying due to the flagrant waste of resources by the sinful acts of modern man. Most of us still refuse to see the dying universe and continue to amuse ourselves in the illusory funfair by disregarding the prophetic signs given by Thoreau-like intellectuals over the prospect of calamities on the planet as a consequence of science's impulsive actions. If Thoreau-like writers' prophetic visions in their post/apocalyptic works had been valued as much as their stylistic aspects, the term "Anthropocene" would not perhaps sound so apocalyptic now. However, as Haraway et al. argue, we ought not to be desperate at all. This tragedy "holds an odd, even schizophrenic promise; namely the promise of scientific renewal and insight."[3] There is still hope to recover the unintended/intended damages caused to the planet by science as long as we continue our consciousness-raising efforts particularly in stopping self-destructive games on an already dying planet. To this end, this chapter firstly discusses the extent of the human-made impact upon nature. This impact has even reached the point of causing a geological shift. Then, the focus will shift to the discussion of Thoreau's *Walden* as a prophetic work. This chapter humbly aspires to show how Thoreau's foresight in *Walden* attempts to awaken seemingly civilized minds over the reconsideration of humans' dangerous intrusion into the natural functioning of the universe.

---

[2] William Wordsworth, *The Prelude: Or, Growth of a Poet's Mind; an Autobiographical Poem*, vii 645-49.
[3] Donna Haraway, et al., "Anthropologists are talking–about the Anthropocene" *Ethnos* no. 81.3 (2016): 535.

Anthropo[sin]: The Age of the Sinful Man

A recently published article in *Nature* carried out by a group of researchers led by Dr. Milo has put forward some significant and bitter results on how human-made objects will likely outweigh all living beings on Earth. In his interview with BBC news, Milo says, "the significance is symbolic in the sense that it tells us something about the major role that humanity now plays in shaping the world and the state of the Earth around us."[4] Though the figures seem to be symbolic, they might be interpreted as evidence of the pervasive actions of humankind on Earth. The perilous encroachment of mankind on the natural workings of the cosmos has had a significant impact on the world's biosphere to the point that our geological epoch has been called the Anthropocene, a term popularized by Eugene Stoermer and Paul Crutzen:

> The term Anthropocene . . . suggests that the Earth has now left its natural geological epoch, the present interglacial state called the Holocene. Human activities have become so pervasive and profound that they rival the great forces of Nature and are pushing the Earth into planetary terra incognita.[5]

Although the term Anthropocene had been used in published articles before, it is Crutzen that made it quite popular when he interrupted his fellow scientists and told them to stop calling the geological epoch the Holocene at a meeting of the International Geosphere-Biosphere Program in Cuernavaca, Mexico. He says: We're not in the Holocene any more. We're in the . . . the . . . the . . . (searching for the right word) . . . the Anthropocene!".[6] Crutzen asserts that the transformative acts of human societies on Earth have become a massive force since the Industrial Revolution. Therefore, the steam engine, as they put forth, symbolically represents the start of the Anthropocene.

---

[4] Helen Briggs, "Human-made objects to outweigh the living things". Accessed on March, 10, 2021. https://www.bbc.com/news/science-environment-55239668.
[5] Will Steffen, Paul J. Crutzen and John R. McNeill, "The Anthropocene: Are Humans Now Overwhelming the Great Forces of Nature?" *Ambio*, no. 38 (2007): 614.
[6] Angus Ian, *Facing the Anthropocene: Fossil capitalism and the crisis of the earth system* (NYU Press, 2016), 28.

Like some other humanists, Jamieson, an important scholar, believes that this term may sound the golden age of humankind.[7] However, if we pun on the word, "cene", it is likely to be associated with the era of the sinful man, "anthropo[sin]". In the tragic and intentionally/unintentionally destructive actions of humans on Earth, there is no dignity, no fame but only atrocity. The shaping force of humans on the natural cycle of the ecosystem inevitably results in what Bennett, a critic, calls a paradigm shift.[8] Now, Earth is on fire. It suffers from the negative consequences of the inconsiderate acts of humans due to their unappeasable desire to dominate, and to colonize the earth. These may include climate change, environmental pollution, the melting glaciers of the Arctic, rising sea level, extreme weather conditions, the piled-up toxic wastes on the entire planet and new pop-up viruses like Covid-19 that have transvalued almost all of the socially constructed values. Buxton and Hayes, academic fighters for environmental justice, question the reasons for not taking the matter too seriously.[9] Instone, a scholar who has concerns for life in the Anthropocene, may possibly give the best response to this question by linking Beck's notion of "risk society" to humankind's disregard of the alarming cries of a dying nature as a result of the harm that it has inflicted on nature. To Beck, a significant social scientist, the insurance facilities in almost every aspect of life normalize the risk factor, so people get accustomed to living through the risks.[10] Instone argues that people are at ease with their actions, denying their potentially harmful effects on nature as if everything in nature is insurable.[11] Therefore, Instone is right in blaming them for amnesia.

Bracke, an important researcher like Instone, refers to this amnesiac state as "cognitive dissonance".[12] She asserts that people have a temporary consciousness of climate change. They continue to live their lives as if nothing were happening. This amnesiac attitude is not a new

---

[7] Dale Jamiesen, "The Anthropocene: Love it or Leave it". In *The Routledge Companion to the Environmental Humanities*, edited by Ursula K. Heise, Jon Christensen, and Michelle Niemann (Taylor & Francis, 2017), 14.
[8] Jill Bennett, *Living in the Anthropocene* (Hatje Cantz, 2011), 6.
[9] Nick Buxton and Ben Hayes, "Introduction: Security for Whom in a Time of Climate Crisis?" *The Secure and the Dispossessed* (Pluto Press, 2016), 7.
[10] U. Beck, *Risk Society: Towards a New Modernity* (London: Sage, 1992), 22.
[11] Lesley Instone, "Risking Attachment in the Anthropocene". In *Manifesto for Living in the Anthropocene*, edited by Katherine Gibson, Deborah Bird Rose, and Ruth Fincher (Punctum Books, 2015), 30.
[12] Astrid Bracke, *Climate Crisis and the 21st-Century British Novel* (Bloomsbury Publishing, 2017), 3.

agenda. Despite their diametrically opposed styles, literary figures have foreshadowed these potential disasters (which we are currently experiencing) with their prescient pens in their works, but they have not been taken into account. Their works have been read as pleasurable fictions that are aesthetically styled. The apocalyptic aspect of the fiction has long been ignored or of little value. Despite the presence of horrifying events in post/apocalyptic fiction, the readers often dismiss the likelihood of these catastrophic scenarios happening. People have a tendency to easily miss things. Post-apocalyptic literature has staged numerous dystopian nightmares for the readers. One may never forget Thoreau's apocalyptic warning of the mechanical conversion of people, Mary Shelley's grotesque figure as a result of the scientific creation, and Huxley's scathing painting of the seemingly ideal nature in the cinematographic world in *Brave New World*. There are many other examples to count such as N'gugi's metaphoric comparison of the first-coming train into Nairobi with an iron snake ready to devour/colonize nature, Eliot's comparison of the capitalist world to a dog that feeds on the bodies in *Waste Land*, or Carson's apocalyptic manifesto on the use of pesticides in *Silent Spring*. Thoreau-like figures have always been present on the stage of human history with their forebodings and have shown us how to avoid the impending dangers and pitfalls through their witty metaphors if read seriously. With this purpose in mind, the following section will conduct a multilateral analysis of *Walden* in order to uncover Thoreau's long ignored apocalyptic vision.

Thoreau's Prophetic Vision in *Walden*

Thoreau's *Walden* can be read as a warning for humankind against the impending threats of the industrial society to the biosphere. The prophetic quality of his writing underpins his account of experiences that he had in a detached hut in the woods. In one of his attributions to Thoreau, Seybold describes him as a poet and prophet who "discerned the open secrets of the universe".[13] It may not be wrong to claim that Thoreau is rewarded with prophetic vision only after blinding his public eyes. In other words, his detached and oppositional stance toward the current events in his time seems to strengthen his prophetic vision. From his early life on, his protests against social injustices enabled him to see with the eye of the heart. As W.S. Mervin mentions in his introduction to *Walden*,

---

[13] Ethel Seybold, "Thoreau: The Quest and the Classics." In *Henry David Thoreau*, edited by Harold Bloom (Infobase Publishing, 2007), 20.

his refusal to pay for the diploma at Harvard, the termination of his contract for his first teaching post due to his outrage at the inequity, and his refusal to pay taxes in order to oppose war and slavery are some of the reasons that have awakened him.[14]

Thoreau's transcendentalist views as well as his life experiences help him move beyond the material boundaries of the existing system. As a significant critic on Thoreau, Robinson stresses the fact that both Thoreau's diverse readings including Plato's works and Hindu scriptures and his intellectual meetings with Emerson devote him to the transcendental life, which has become more significant for him as the slavery issue worsens.[15] Upon a question directed at him by the secretary of a scientific circle called *The American Association for the Advancement of Science*, Thoreau calls himself "a mystic–a transcendentalist, a natural philosopher to boot".[16] Thoreau's transcendental ideas help him see the intangible force that spiritually connects all the most disparate objects in nature. Similarly, Robinson observes that Thoreau's Self-awakening by his discovery of the symbiotic energy in nature reflects a new insight into the organic relationship between humans and nature in *Walden*.[17] It is an undeniable fact that his transcendentalist beliefs endow him with prophetic sight. However, his vision becomes more tragic as it agonizingly exacerbates his concerns about the impending threats of industrialization and the eventual extinction of nature's life-giving resources. His worries and anxieties over the ignorance and doziness of humans in *Walden* appear to impose on him the task of raising consciousness in his fellow people.

Thoreau's transcendental view as well as his life experiences clearly distance him from the illusory and destructive life created by the rising industries. His self-liberating acts not only open his eyes but also make him a missionary to rouse others from their slumber. Thus, presenting his own account of the life he spent in the woods for two years, Thoreau attempts to hold a mirror to the wrongdoings of the inhabitants in their comfortably founded towns/cities, which would ultimately become their own prison and make them captives or "serfs of the soil":

---

[14] Henry David Thoreau, *Walden or, Life in the Woods and on the Duty of Civil Disobedience* (1960), 3.
[15] David M. Robinson, "Living Poetry". In *Henry David Thoreau,* edited by Harold Bloom (Infobase Publishing, 2007), 128.
[16] Nina Baym, "Thoreau's View of Science." *Journal of the History of Ideas*, 26.2 (1965): 233.
[17] Robinson, "Living Poetry", 128.

Who made them serfs of the soil? Why should they eat their sixty acres, when man is condemned to eat only his peck of dirt? Why should they begin digging their graves as soon as they are born?[18]

Walden begins with Thoreau's critique of the townsmen who "labored under a mistake".[19] In Marxist terms, they become self-alienated figures since they cut off their organic connections to their true nature, thus making themselves the willing prisoners of property. These self-alienated figures are, as Thoreau asserts, "frittered away with detail".[20] In the industrialized community, the people's preoccupation with details and their indulgence in the artificial life are the potential outcomes of the rush of life and the newly developed social relations and dynamics. The stunning pace of life rips them away from the organic ties with nature. As Thoreau points out, the people's mental and physical bondage to the new dynamics of the society within the new network of the master-slave relationship blinds them so they do not realize the "unprofaned part of the universe"[21] right under their very nose. Thoreau finds the voluntary bondage of the townsmen in "gold or silver fetters"[22] pathetic and alerts the readers about the dangerous and irreversible effect of this masochistic pleasure on nature.

Unfortunately, nature is bleeding now due to people's disregard of such ominous warnings. One of NASA's most recent images exemplifies the possible outcomes of such ignored prophetic visions.[23] It starkly shows how humans continue to wreck the ecosystem of their only habitat. The gold pits and dirt from illicit mining wash into the rivers of the Peruvian Amazon, poisoning the water and killing the different species living in the rivers. Moreover, according to some experts, the increase in deforestation has reached an alarming level due to mining for more precious metals in different parts of the world. In *Walden*, Thoreau issues some apocalyptic warnings against the possible devastation of nature, observing how his fellow countrymen are "so occupied with the factitious cares and superfluously coarse labors of life".[24] This bus[y]ness of life, as

---

[18] Henry David Thoreau, *Walden*, edited by J. Lydon Shanley (Princeton and Oxford: Princeton University Press, 2004), 5.
[19] Thoreau, *Walden*, 5.
[20] Thoreau, *Walden*, 91.
[21] Thoreau, *Walden*, 88.
[22] Thoreau, *Walden*, 16.
[23] Brandon Specktor, "Rivers of gold rush through the Peruvian Amazon in stunning NASA photo". Accessed on March, 2, 2021. https://www.livescience.com/nasa-peru-amazon-gold-pits-photo.html.
[24] Thoreau, *Walden*, 6.

Thoreau asserts, would possibly eventually turn humans into "a machine".[25] On the 150th anniversary edition of *Walden*, Updike, in his introduction, stresses the crushing effect of the mechanical world where "the steam engine [was] the technological ultimate".[26]

The mechanical aspect of the industrial life is at full speed during Thoreau's stay in Walden. He is an eyewitness to how his fellow townsmen have become the cogs in various parts of the big machine including the railroads, factories, and many other forms of the machine industry. They are trapped in this mechanical cage but seem to have no complaints at all. Thoreau refers to them as "sleepers" that have been enchanted by the mechanical industry's artificial luxuries. One of these "sleepers" worth mentioning is John Field, an Irish immigrant. His desire to hunt for luxuries gives us the mirror image of the newly born industrial society. Thoreau thinks of him as a wasted life lost in the deep illusion. In one of their conversations, he boasts of the railroad with these words; "is not this railroad which we have built a good thing?".[27] This drives Thoreau to question the huge rift between him and his fellow Concord people.

Thoreau perceives the railroad network as an irritating metaphor beyond its function. He sees it as a threshold between sacred nature and the poisonous culture industry dominated by the machine. However, people like John Field enjoy riding on the train. Even the buzzing sound is repulsive to Thoreau. For him, the railroad symbolically represents the boundaries of the rising capital industry through which an artificial life is created. The glittering luxuries of this artificiality lull people to sleep, causing them to miss out on nature's true riches. When reading carefully, Thoreau's apocalyptic vision is evident again in his echoing complaints about the railroad. He says, "we do not ride on the railroad; it rides upon us".[28] It becomes a gruelling burden that his fellow people gladly bear without protest. He rightly blames his people for slumbering since they are only concerned with their welfare for the instantaneous joys and delights. However, Thoreau's genius can be seen in his treatment of the subject beyond its utilitarian purpose. He takes the issue seriously, seeing it as a destructive force that, in the long run, will pollute nature's purity. Thoreau justifies the benevolent intent that lies in science and the mechanical rationality for the welfare of humanity in one or two of his discussions, and he admits enjoying it to some degree. John Hildebidle, an important

---

[25] Thoreau, *Walden*, 92.
[26] Thoreau, *Walden*, XVI.
[27] Thoreau, *Walden*, 54.
[28] Thoreau, *Walden*, 92.

scholar on Thoreau, also features this point in his article claiming that Thoreau may have benefited from scientific methods to achieve his transcendental tasks in nature.[29] However, Thoreau stresses that the seemingly superior mind adopts a one-dimensional approach no matter how innocuous it is. This unilateral strategy will have some dangerous impacts on the evolutionary functioning of nature if intentional/unintentional.

Given the impending threats of the newly introduced industrial and mechanical explorations on nature's delicate equilibrium, Thoreau's advice to his readers is quite clear: "Simplicity!" It is the way to lighten one's burden by offloading the unnecessary details of the material world. According to Thoreau, what is worse is that people become "the slave drivers" of themselves. He sees the simple life as essential to Self-redemption since it awakens their souls. This claim is supported by his removal of three pieces of limestone:

> I had three pieces of limestone on my desk, but I was terrified to find that they required to be dusted daily, when the furniture of my mind was all undusted still, and threw them out the window in disgust. How, then, could I have a furnished house? I would rather sit in the open air, for no dust gathers on the grass, unless where man has broken ground.[30]

Thoreau further argues that nature is a living organism and all living organisms are part of nature. His observation of the newly emerged forms in the muddy water near the railroad gives him some solace in the midst of his worries on dying nature. Nature's durability as a living entity against the invading force of the mechanical industry symbolically represented by the railroad keeps him from succumbing to utter desperation. Thoreau interprets nature's pushback as a second chance for mankind to rethink their reckless and destructive acts. As a result, he urges his fellow readers to break free from civilization's artificial shell, which has separated them from their real existence. In the last pages of *Walden*, his advice to his fellow citizens points towards the path to Self-exploration in nature:

> Direct your eye sight inward, and you'll find
> A thousand regions in your mind
> Yet undiscovered. Travel them, and be
> Expert in home-cosmography.[31]

---

[29] John Hildedible, "Naturalizing Eden: Science and Sainthood in Walden". In *Henry David Thoreau,* edited by Harold Bloom (Infobase Publishing, 2007), 35.
[30] Thoreau, *Walden*, 36.
[31] Thoreau, *Walden*, 320.

He exhorts his fellow Concord people to look inward. He argues that they should live in harmony with nature and be "expert in home cosmography" in order to do so. Only in this manner will mankind find nobility in the savage and raw life.

Conclusion

> Is it impossible to combine the hardiness of these savages with the intellectualness of the civilized man?[32]

Thoreau questions the possibility of merging one's raw nature with the true nobility of the mind not poisoned by the newly emerged dynamics accompanied by industrialization. He is well aware of the fact that industrialization imbues people with false hopes, thereby sending them drifting onto a one-dimensional path, if we talk more specifically, onto the path of self-alienation. These self-alienated figures, overwhelmed by the languor of the industrial society's luxuries have almost lost their organic connection to their true nature where, as Thoreau asserts, one can reach the nobility of the savage life.

Contemporary readers may well justify Thoreau's prophetic vision in *Walden* since his fears over the destruction of nature caused by humans' greed and uncontrollable desire to dominate have become our bitter truths, the repercussions of which we have had to bear. However, we still keep disregarding such apocalyptic visions. Humans' self-masochistic pleasure seems to be never ending. The melting glaciers, the vast amount of floating garbage in the Pacific Ocean, greenhouse gas emissions into the atmosphere, climate change, the increasing heat in the Earth's crust and a slew of other catastrophic events all point to humanity's folly. The question is, as Pope well phrases, why "[do humans] rush in where the angels fear to tread"?[33] Nature is not an eternal supplier for such avarice. Now, it is giving alarming cries with all of these disasters as a result of our dangerous games that we have been playing on our home planet. Nature gives the harsh reactions to these destructive games. For example, Covid-19 is one of them. For humanity, it might be seen as a deadly virus that destroys everything special to the human realm. However, it seems that it is an antivirus that fights against real infection in the universe: the android humans who have cut off their organic connections with the universe. Humans become the deadly, uncontrolled cancerous cells that would

---

[32] Thoreau, *Walden*, 13.
[33] Alexander Pope, *Essay on Criticism* (CUP archive, 1908), line 625.

eventually destroy their own habitat. What is more pathetic is not what they become but their joyous celebration of it! Rather than healing the deep wounds they have inflicted on Earth, they, the cancerous humans, have long conceded Earth's demise and have begun a seemingly progressive attempt to terraform planets like Mars through their expansive policies. It proves how perverse they are at not restricting their destruction to this world but to spread the cancer deep into the universe. Thoreau, no doubt, prophetically draws attention to the pitfalls of humans' mechanical conversion in *Walden* and makes some serious exhortations to future generations to abstain from this disastrous course before it is too late. However, it seems that they are still unheard. What humans do is have a bitter and tragic glimpse of his prophetic cries in *Walden* since they keep watching how time has justified them one by one.

References

Angus, Ian. 2016. *Facing the Anthropocene: Fossil capitalism and the crisis of the earth system*. NYU Press.
Baym, Nina. 1965. "Thoreau's View of Science." *Journal of the History of Ideas*, 26.2: 221-234.
Beck, U. 1992. *Risk Society: Towards a New Modernity*. London: Sage.
Bennett, Jill. 2011. *Living in the Anthropocene*. Hatje Cantz.
Bracke, Astrid. 2017. *Climate Crisis and the 21st-Century British Novel*. Bloomsbury Publishing.
Briggs, Helen. "Human-made objects to outweigh the living things". accessed on March 10, 2021. https://www.bbc.com/news/science-environment-55239668.
Buxton, Nick, and Ben Hayes. 2016. "Introduction: Security for Whom in a Time of Climate Crisis?" In *The Secure and the Dispossessed*, 1-19. Pluto Press.
Haraway, Donna, et al. 2016. "Anthropologists are talking–about the Anthropocene." *Ethnos* 81.3: 535-564. DOI:10.1080/00141844.2015.1105838.
Hildedible, John. 2007. "Naturalizing Eden: Science and Sainthood in Walden". In *Henry David Thoreau,* edited by Harold Bloom, 35-61. Infobase Publishing.
Instone, Lesley. 2015. "Risking Attachment in the Anthropocene". In *Manifesto for Living in the Anthropocene*, edited by Katherine Gibson, Deborah Bird Rose, and Ruth Fincher, 29-36. Punctum books.
Jamiesen, Dale. 2017. "The Anthropocene: Love it or Leave it". *The Routledge Companion to the Environmental Humanities*, edited by

Ursula K. Heise, Jon Christensen, and Michelle Niemann, 13-20. Taylor & Francis.

Pope, Alexander. 1908. *Essay on Criticism*. CUP archive.

Robinson, David, M. 2007. "Living Poetry". In *Henry David Thoreau,* edited by Harold Bloom, 127-151. Infobase Publishing.

Seybold, Ethel. 2007. "Thoreau: The Quest and the Classics." In *Henry David Thoreau,* edited by Harold Bloom, 13-34. Infobase Publishing.

Specktor, Brandon. "Rivers of gold rush through the Peruvian Amazon in stunning NASA photo". Accessed on March 2, 2021. https://www.livescience.com/nasa-peru-amazon-gold-pits-photo.html.

Thoreau, Henry David. 1960. *Walden or, Life in the Woods and on the Duty of Civil Disobedience.*

Thoreau, Henry David. 2004. *Walden*, edited by J. Lydon Shanley. Princeton and Oxford: Princeton University Press.

Will Steffen, Paul J. Crutzen and John R. McNeill. 2007. "The Anthropocene: Are Humans Now Overwhelming the Great Forces of Nature?" *Ambio* 38: 614-21.

Wordsworth, William. 1888. *The Prelude: Or, Growth of a Poet's Mind; an Autobiographical Poem.* D.C. Heath.

CHAPTER TWO

PRE-HISTORIES OF CLI-FI: THE AGENTIAL REALITY OF RICHARD JEFFERIES'S *AFTER LONDON*

ADRIAN TAIT

Introduction

As we enter what is now widely regarded as the Anthropocene epoch, there is a growing interest in climate change fiction, or "cli-fi,"[1] and its potential to alert readers to the potential consequences of a changed climate.[2] Many of these fictional narratives have used the possibility of rising sea levels to dramatize the effects of climate change. As Adam Trexler notes, "[f]loods offer a rich, literary means of rendering climate change in a local place, as a tangible concrete effect."[3] Moreover, add Andrew Milner and J. R. Burgmann, the flood is a trope with "a deep history in the Western mythos."[4] It is not therefore surprising that, "[o]ver the last forty years," the flood has become "the dominant literary strategy for locating climate change."[5]

---

[1] Adeline Johns-Putra, "Climate change in literature and literary studies: from cli-fi, climate change theatre and ecopoetry to ecocriticism and climate change criticism," *WIREs Clim Change* 7 (2016): 267,
https://onlinelibrary.wiley.com/doi/abs/10.1002/wcc.385.
[2] For a discussion of that potential, see Matthew Schneider-Mayerson, "The Influence of Climate Fiction: An Empirical Survey of Readers," *Environmental Humanities* 10, no. 2 (2018): 473-500.
[3] Adam Trexler, *Anthropocene Fictions: The Novel in a Time of Climate Change* (London: University of Virginia Press, 2015), 83.
[4] Andrew Milner and J. R Burgmann. "A Short Pre-History of Climate Fiction," *Extrapolation* 59, no. 1 (2018): 6.
[5] Trexler, *Anthropocene Fictions*, 82.

As critics have also suggested, these flood narratives have a pre-history.[6] A whole series of "deluge" novels "act as precursors" to the contemporary cli-fi flood narrative,[7] including works such as J. G. Ballard's *The Drowned World* (1962). "Years before the first novel about human-induced global warming," Trexler argues, novels such as this "provided a strikingly stable archetype for subsequent [climate change] fiction."[8] Yet Ballard's novel was by no means the first to describe the impact of rising sea levels on a recognisably modern, developed world, or establish the parameters for its depiction. In *After London; or Wild England*, first published in 1885, Richard Jefferies (1848-1887) described a near future in which a now fragmented and barbaric society struggles to survive the aftermath of a mysterious flood.[9] Much of southern England is now underwater, and London itself has disappeared, lost in a poisonous swamp. In this brutish world, it is not now known what, exactly, prompted the flood, and there is only the sketchiest understanding of how a civilised and sophisticated society came undone so quickly; all that seems certain is that, in spite of its many marvels, the modern world lacked the resilience to ensure its continuance.

As this brief outline suggests, Jefferies's remarkable novel itself provides a compelling template for depicting a climate-changed world. As I have argued elsewhere,[10] many of the tropes that characterise *After London*, such as the fragility of modern society, reappear in much more recent novels such as Robert Harris's *The Second Sleep* (2019). However, one of the most interesting aspects of the novel – its detailed account of the way that the flood overwhelms a modern city's infrastructure – has so far been given little attention, perhaps because Jefferies's portrayal of it can so easily be read in symbolic or mythic rather than literal terms, or because the

---

[6] Trexler, *Anthropocene Fictions*, 84-87; Milner and Burgmann, "A Short Pre-History," 5.
[7] Trexler, *Anthropocene Fictions*, 86.
[8] Trexler, *Anthropocene Fictions*, 87.
[9] For an excellent introduction to the novel, readers may wish to refer to Mark Frost's recent scholarly edition of *After London*. Mark Frost, introduction to Richard Jefferies, *After London; or Wild England*, ed. Mark Frost (Edinburgh: Edinburgh University Press, 2017), vii-lii.
[10] Adrian Tait, "Environmental Crisis, Cli-fi, and the Fate of Humankind in Richard Jefferies' *After London* and Robert Harris' *The Second Sleep*," *Exchanges: The Interdisciplinary Research Journal, Special Issue: Climate Fiction* 8, no. 2 (2021): 69-83, https://doi.org/10.31273/eirj.v8i2.

novel itself has been banished to what Amitav Ghosh has described as "the outhouses [of] science fiction and fantasy."[11]

Ghosh's related point is that serious, realist fiction is simply unable to rise to the challenge of depicting the impact of climate change, such as "catastrophic aquatic events."[12] Yet Ghosh is, Ursula Kluwick maintains, "too hasty in his dismissal of the achievements of 'serious fiction.'"[13] What is needed is a new way of reading attuned to some of the subtler ways in which novels like Jefferies's have engaged with "aquatic events" and the "nonhuman agency" they embody[14]. What is required, in other words, are "more material interpretations"[15] of "water fictions" that resist the temptation to regard their account of an agential and processual reality as simply figurative or mythological or symbolic or (after Ghosh, and his problematic dismissal of science fiction) other worldly;[16] what is required is a new, materialist (or new materialist) willingness to read these novels not just as realist, but as more-than-realist, and explore the (perhaps forgotten or overlooked) socio-historical context in which they were written. As I argue in this chapter, Jefferies's account of a sunken and swamp-bound London should be read not as some sweeping symbolic dismissal of modern society or as an otherworldly instance of speculative fiction, but as a nuanced and exacting response to contemporary anxieties about the vulnerabilities of the modern megalopolis; and as our own urbanised society itself faces the problem of a warmed and expanding ocean – and the possibility of our own, anthropogenic deluge – those anxieties are (or should be) still with us.

After London as Cli-fi

After London is divided into two, uneven parts. In the first, an unnamed scholar recounts what has happened in the years since civilisation collapsed, a collapse that has allowed a self-willed, nonhuman world to regenerate, whilst human society has itself fragmented into warring factions. In the second part, the novel's putative hero, Felix, embarks on a journey of (self-

---

[11] Amitav Ghosh, *The Great Derangement: Climate Change and the Unthinkable* (London: University of Chicago Press, 2016), 66.
[12] Ursula Kluwick, "The global deluge: floods, diluvian imagery, and aquatic language in Amitav Ghosh's *The Hungry Tide* and *Gun Island*," *Green Letters – Studies in Ecocriticism; Special Issue: Waters Rising* 24, no. 1 (2020): 67.
[13] Kluwick, "The global deluge," 67.
[14] Kluwick, "The global deluge," 67.
[15] Trexler, *Anthropocene Fictions*, 87.
[16] Kluwick, "The global deluge," 67; Ghosh, *The Great Derangement*, 72.

)discovery. That quest leads him into poisonous swamp marking the location of what was once an empire's capital. Here, he encounters islands wrapped in fog, where fire flickers over the land and there are everywhere ashy reminders of a great city:

> Presently a white object appeared ahead; and on coming to it, he found it was a wall, white as snow […] He touched it, when the wall fell immediately with a crushing sound as if pulverised […] Whether the walls had been of brick or stone or other material he could not tell; they were now like salt.[17]

What the purifying flood has not engulfed, fire has reduced to "a white power."[18]

Almost inevitably, these remnants conjure up the Biblical story of Sodom's fall, consumed by fire, and of the fate of Lot's wife, turned to a pillar of salt when she presumes to look back on the city's destruction. The symbolism of the scene is difficult to ignore: London has suffered the same fate as its Biblical forebear, and for the same reasons: it is, to paraphrase Darko Suvin, a site of decadent wealth, luxury, and pride, richly deserving its own destruction, and Jefferies describes its obliteration with "equanimity and indeed with relief."[19] Perhaps he even does so with relish, hating the city because, as David Garnett argued, it "robbed him of his health."[20] According to this reading, the surreal, nightmare-like quality of Felix's encounter with the ruined city – "[h]is brain became unsteady, and flickering things moved about him" – simply underlines its essential unreality.[21] As John Fowles remarked in his own discussion of this haunting vision of "near-madness,'" "[o]ne may argue over the physical plausibility of the world Jefferies envisages; but not, I think, over its metaphorical power."[22] What we should not do, in other words, is take these passages literally; they represent Jefferies's symbolic stand against the "bourgeois

---

[17] Richard Jefferies, *After London*; or Wild England, ed. Mark Frost (Edinburgh: Edinburgh University Press, 2017), 164-5.
[18] Jefferies, *After London*, 165.
[19] Darko Suvin, *Victorian Science Fiction in the UK: The Discourses of Knowledge and of Power* (Boston: G. K. Hall, 1983), 373.
[20] David Garnett, introduction to Richard Jefferies, *After London and Amaryllis at the Fair* (London: J. M. Dent & Sons, 1939), ix.
[21] Jefferies, *After London*, 164.
[22] John Fowles, introduction to Richard Jefferies, *After London, or Wild England* (Oxford: Oxford University Press, 1980), xix.

civilization" of his own day, and as Felix moves through the ruins, the material simply denotes the abstract.[23]

But what if we reverse this reading? What if we regard London's destruction not as a latter-day fable, without grounding in fact or possibility, nor even as a "fictional hypothesis,"[24] but as the logical working-out of genuine anxieties? "All these matters are purposely dealt with in minute detail so that they may appear actual realities," Jefferies wrote in a letter referring to the novel; "[t]he state of the site of London is fully described" (April, 1884).[25] These are not the words of a writer intent on the fabular, but rather, those of a writer insisting that his work is in the realist mode. The related question is whether in constructing the appearance of "actual realities" Jefferies is simply using this detailed picture of the material world as part of a Barthian "reality effect" (the detail is in itself useless, and "in its very superfluity resists serving narrative meaning and so simply announces the reality of the world represented by the text").[26] To the contrary, and as I explain in the next part of the chapter, a careful reading of the novel's socio-historical and intellectual context indicates that the material world on which Jefferies focuses such careful discussion is not part of an "effect," but is an integral and important part of a novel which is everywhere concerned to reassert the dynamism of the nonhuman and more-than-human worlds.

To make this point is, of course, to adopt a new materialist perspective, with its emphasis on the constitutive entanglements between people and place, culture and nature, the role of reality as itself agential, and on discourse as in fact part of an active material-discursive formation, with its own constitutive role to play.[27] As the proponents of new (or vital) materialism have suggested, things have power; they are active, and often, they act in ways that unpredictable and unwanted. Furthermore, argues Jane

---

[23] Suvin, *Victorian Science Fiction*, 373.
[24] W. J. Keith, *Richard Jefferies: A Critical Study* (London: Oxford University Press, 1965), 117.
[25] Quoted in Walter Besant, *The Eulogy of Richard Jefferies* (New York: Longmans, Green & Co, 1888), 210.
[26] Caroline Levine, "Victorian realism," in *The Cambridge Companion to the Victorian Novel*, ed. Deidre David (Cambridge: Cambridge University Press, 2012), 93.
[27] For a fuller account of these aspects of new materialist thinking, see Karen Barad, *Meeting the Universe Halfway: Quantum Physics and the Entanglement of Matter and Meaning* (London: Duke University Press, 2007), 32-36; Stacey Alaimo, *Bodily Natures: science, environment, and the material self* (Bloomington: Indiana University Press, 2010), 1-3; and Jane Bennett, *Vibrant Matter: a political ecology of things* (London: Duke University Press, 2010), vii-xi.

Bennett, "a vital materiality can never really be thrown 'away,' for it continues its activities even as discarded or unwanted commodity."[28] Rubbish, trash, and waste have a way of rebounding on us, as Jefferies's novel emphasises in the most graphic ways.

Before engaging with this aspect of *After London*, a novel in which waste and pollution feature as significant factors in civilisation's destruction, the first step in this new materialist rereading is to address the problematic question of what, exactly, acts as the catalyst for its downfall. The unnamed scholar of the novel's first part provides an account of how this fictional future society came into being, an account pieced together from folk memory, "written notes,"[29] and a handful of "records," most of which "were destroyed in the conflagrations [...] which consumed the towns."[30] Although the scholar is intent on presenting the most accurate account of "the event"[31] and its aftermath, however, he openly admits that "nothing is certain and everything confused."[32] In the novel's second part, the narrative is taken up by a third-person narrator, who relates Felix's discoveries, but that narrator is also at a loss to know exactly what has happened, or why.

Both narrators therefore accept that the "ultimate truth" may never be known.[33] Nonetheless, it is possible to piece together hints and clues. The source of the disaster was, it seems, a rise in sea levels, inundating cities such as London; when sea levels fell back, they left behind sand and silt and debris, blocking the Thames (along with other rivers), and creating the vast lake that now occupies much of southern England.[34] Amidst the chaos, food

---

[28] Bennett, *Vibrant Matter*, 6.
[29] Jefferies, *After London*, 14.
[30] Jefferies, *After London*, 4.
[31] Jefferies, *After London*, 15.
[32] Jefferies, *After London*, 14.
[33] Jefferies, *After London*, 14.
[34] Jefferies, *After London*, 14, 30. This puts to one side the question of what caused the sea level rise in the first place. Attuned to the discourse of climate change, we are today alert to the fact of rising sea levels, and forewarned that, since cities like London and New York lie so close to sea level, they are at immediate risk. Whether or not this is what Jefferies had in mind is another matter. A "greenhouse effect" was first proposed by Jean-Baptiste Joseph Fourier in 1824, but the theoretical basis for global warming was not set out until the 1890s; see Mike Hulme, *Why We Disagree About Climate Change: Understanding Controversy, Inaction and Opportunity* (Cambridge: Cambridge University Press, 2009), 42, 46. Anthropogenic climate change is not, therefore, a likely influence on the plot of *After London*. There is, however, an alternative, non-anthropogenic explanation for a changed climate that might have had an influence on the narrative. Briefly, the narrator toys with the

supplies were interrupted, and the apparatus of a complex modern society broke down completely.

Why was this breakdown irrevocable? Here, two possible interpretations open up. The first is that those with the wherewithal fled the country;[35] when "the cunning artificers of the cities all departed," "the secrets of their sciences" were soon lost, and "everything fell quickly into barbarism."[36] Yet "nothing has ever been heard" of those "that left the country,"[37] opening up the possibility of a second explanation: that, in fact, no one left, and the multitudes who filled the cities, rich and poor alike, died there, an alternative explanation corroborated by stories of the other "ancient cities" whose sites – some now "lost in the forest," others covered by swamp – "are avoided" because they cause "ague or fever."[38]

---

possibility that "the earth, from some attractive power exercised by the passage of an enormous dark body through space, became tilted or inclined to its orbit more than before" (Jefferies, *After London*, 14). The tilt of the earth's axis creates the seasons, and itself alters over a cycle of 41,000 years. The more the earth tilts away from the sun, the less sunlight the poles receive during their winters, creating relatively more ice. The reverse is also true: a tilt towards the sun creates a positive feedback mechanism that melts the ice-caps and causes sea levels to rise. This connection was first made by a Victorian, James Croll (1821-1890). Croll argued that "the 'true cosmical cause' of climate change 'must be sought for in relations of our earth to the sun;'" see James Rodger Fleming, "James Croll in Context: The Encounter between Climate Dynamics and Geology in the Second Half of the Nineteenth Century," *History of Meteorology* 3 (2006): 43. Whilst the mechanism that Croll described is simple enough, it was not until Milutin Milanković elaborated Croll's ideas in 1924 that they were fully accepted by the scientific community (Hulme, *Why We Disagree*, 42). It has, furthermore, taken several more decades of research to separate out anthropogenic effects from non-anthropogenic explanations for climate change, and develop a serviceable case for human-induced global warming. Nevertheless, Croll's book on *Climate and Time* was published (1875) around the time when Jefferies was first sketching out his ideas for *After London*, and it is possible that Jefferies had non-anthropogenic climate change in mind when he wrote the novel.

[35] Jefferies, *After London*, 15.
[36] Jefferies, *After London*, 17, 16, 17. The world whose downfall the scholar describes is recognisably modern, and to the scholar, marvellous: with wireless, railways, and steam-ship, it spanned the globe and raised buildings 'to the skies." Jefferies, *After London*, 16-17, 16.
[37] Jefferies, *After London*, 15.
[38] Jefferies, *After London*, 33.

If so, what was it that caused this "plague"?[39] There are two, overlapping accounts, both of which are borne out by Felix's experiences amidst London's ruins. The first is that the multitudes were poisoned by the accidental release of the "strange and unknown chemicals collected by the wonderful people of those times,"[40] chemicals that now form part of the toxic cloud that marks the site of London; this might explain both "the gaseous emanations" that narcotise Felix, and the bizarre atmospheric effects and plays of "Jack-o'-the-lantern […] phosphoric fire" – the "fiery vapours" – that so bewilder him.[41] Mindful of our own experience of living in a risk society imperilled by industrialisation's unpredictable side-effects, this is certainly a fascinating possibility, but the alternative explanation is still more interesting and perhaps more unexpected. Jefferies left behind few clues about the novel he was writing, but in a notebook entry of July, 1884 (written within a few months of the letter, quoted above) he refers to what he describes with horror as "[t]his W. C. [water closet] Century."[42]

> The sewers system and the W. C. water. The ground prepared for the Cholera plague and fever, zymotic [relating to or as if caused by fermentation], killing as many as the plague. The 21 parishes of the Lower Thames Sewage Scheme without any drainage at all. The whole place prepared for disease and pestilence.[43]

One, very obvious explanation suggests itself for these anxieties: that, as Jefferies's reference to "[t]he 21 parishes of the Lower Thames Sewage Scheme" suggests, a modern society had yet to catch up with the sanitation demands of its own rapidly expanding urban basis, a problem that could only be solved – and was already being solved – by the creation of an elaborate sewerage system throughout London: famously, Sir Joseph Bazalgette's vast system of drains and sewers was completed in 1875. Yet Jefferies's notebook entry, with its reference to "[t]he sewers system and the W. C. water," suggests that his anxieties took a slightly different form, whose meaning might now be lost to us, accustomed as we are to the kinds

---

[39] Heidi C. M. Scott, *Chaos and Cosmos: Literary Roots of Modern Ecology in the British Nineteenth Century* (University Park, PA: Pennsylvania State University Press, 2014), 53, 55.
[40] Jefferies, *After London*, 166.
[41] Jefferies, *After London*, 164, 168, 164.
[42] Samuel Looker, ed., *The Notebooks of Richard Jefferies* (London: Grey Walls Press, 1948), 181.
[43] Looker, ed., *The Notebooks*, 181.

of system that Bazalgette pioneered (a system still in operation today beneath London's streets).

As Bill Luckin explains, early nineteenth-century critics of the urban condition "were convinced that overblown new towns and cities would one day sink beneath the weight of their own filth," a "catastrophist vision [that] peaked between the early 1830s and the mid-1850s."[44] The problem was simple, but twofold. Firstly, existing local and voluntary systems for sewage disposal were overwhelmed by the rate and scale of urban expansion. Previously, human waste was emptied into cesspits or pools, which were in turn periodically emptied, and the contents carted away by "night-soil collectors" to provide manure for nearby agriculture; now, these cess-pits and middens were overflowing.[45] Secondly, the water closet was increasingly being used, completely bypassing this longstanding system for recycling excrement. Once the water closet was adopted, house drains (all of which flowed straight into the Thames) were co-opted as a means of carrying away what was now a deluge of sewage, with the inevitable result. In the words of one contemporary, Thomas Cubitt, writing in 1840, "[t]he Thames is now made a great cesspool instead of each person having one of his own."[46] It was not until the Great Stink of 1858 that Parliament voted the money needed to create Bazalgette's networks of sewers, but even Sir Joseph's sewers simply directed the excrement eight miles downstream. Efforts to reproduce the closed system of earlier sewage disposal methods – whereby excrement was recycled as manure – came to nought.[47] The reasons were not just technical or logistical; "[n]othing […] was more loathsome to the Victorian and Edwardian social elites than sewage," and out of sight was out of mind.[48] The question of where the sewage went mattered less than the fact of its removal. Meanwhile, engineers wrestled with "[t]he quest for an 'ultimate sink,'" where this tide of excrement might safely be dumped; but as Luckin asks, "did it, could it ever exist?"[49] By transforming a closed system into an open one, a modern society had created for itself an insoluble problem. The waste it created would simply not go away.

---

[44] Bill Luckin, *Death and Survival in Urban Britain: Disease, Pollution and Environment 1800-1950* (London: Bloomsbury Academic, 2015), 13.
[45] Judith Flanders, *The Victorian City: Everyday Life in Dickens' London* (London: Atlantic, 2012), 206-7.
[46] Quoted in Stephen Halliday, *The Great Stink of London: Sir Joseph Bazalgette and the Cleansing of the Victorian Capital* (Stroud: Sutton Publishing, 2001), 46.
[47] Luckin, *Death and Survival*, 13.
[48] Luckin, *Death and Survival*, 99.
[49] Luckin, *Death and Survival*, 14.

Writ large, this is the problem of modernity itself: it depends on a one-way flow that links resource extraction and consumption to disposal, a flow that is unsustainable both because those resources are not infinite and because the opportunities to dispose of waste-products are themselves not endless. This is precisely the problem that Jefferies dramatizes and transforms into the undoing of society itself: its waste comes back to destroy it. Once again, this is a literal fact rather than metaphorical allusion. Simply, the rising waters have "floated up to the surface the contents of the buried cloacae [from the Latin for sewer]."[50] As the unnamed scholar explains in more depth, the water backed up through "cloacae and drains"[51] – "to fill the underground passages and drains, of which the number and extent was beyond all the power of words to describe" – and burst up, felling houses and returning to the streets the waste products that society triumphantly thought it had dispatched, creating "a vast stagnant swamp, which no man dare enter, since death would be his inevitable fate."[52]

> There exhales from this oozy mass so fatal a vapour that no animal can endure it. The black water bears a greenish-brown floating scum, which for ever bubbles up from the putrid mud of the bottom. […] There are no fishes, neither can eels exist in the mud, not even newts. It is dead.[53]

It is this fatal vapour or "miasma," as the scholar calls it, which marks the site of London as a pestilential, lethal place, and the same is "the case with the lesser cities and towns."[54] Miasma refers to the now discredited notion – then, still a medical orthodoxy[55] – that the foul smells given off by rotting organic matter (such as excrement) might themselves be the cause of disease; now, we better understand that the diseases which beset Victorian society, like cholera, are communicated through the contamination of drinking water by human waste, which might still explain the fevers and plagues that appear to have overwhelmed Jefferies's fictive future. However, and whether or not Jefferies did or could possibly have known as much, the fact also remains that, in slow-moving or static bodies of water like the one that Jefferies describes, natural organic material biodegrades anaerobically, creating something like the zymotic mechanism to which

---

[50] Jefferies, *After London*, 32.
[51] Jefferies, *After London*, 30.
[52] Jefferies, *After London*, 31.
[53] Jefferies, *After London*, 31.
[54] Jefferies, *After London*, 31.
[55] Luckin, *Death and Survival*, 15.

Jefferies refers, above; this might in turn create the gases that imperil those who, like Felix, venture into the heart of what was once London. In other words, the result really could be a poisonous, gaseous swamp of the kind that Jefferies describes.

Thus, Jefferies's remarkable and compelling account of modernity's overthrow hinges on a very different kind of purifying flood, which erases society by returning to it the waste it has pushed away and forgotten: in the present the novel describes, "all the rottenness is there festering under the stagnant water,"[56] sustaining an eco-system inimical to most forms of life, human included. Two points follow. The first relates to the agentiality of the nonhuman world that Jefferies depicts. This is not just a question of the rewilding that the scholar describes with such compelling intensity in the first part of the novel; this much we might expect from a writer like Jefferies, whose eye for nature and natural processes made him one of the most remarkable commentators of his age. Nor is it wholly a question of the flood-waters themselves, compelling as they are as one of "most forceful examples of environmental agency,"[57] and themselves a key reason for positioning Jefferies's novel as part of the pre-history of cli-fi. It is also a question, as Jefferies understood, of the way that modernity generates risks that it seeks not to solve but to elide or escape, not least the risks associated with human waste and industrial pollutants. It is a question, as John Parham has argued, of "material entanglements we would like to ignore, but ought to confront," and in particular, of "the ongoing vitality of things that remain active long after their initial function has been fulfilled – waste products, by-products, garbage, residue, contaminants, run off" – and, we might add, excrement.[58] This matters because, as Parham adds, "nonhuman agency seeps into human society and human being;"[59] inevitably, irrevocably, we are entangled with the world of our own creation, a world that evades the limitations that humankind seeks to impose on it. We are, in other words, a part of (and not apart from) that world, and it registers in what Stacey Alaimo has called our "trans-corporeality," a term that signifies the way these entanglements are felt in the flesh, and are more broadly carried over into the realm of "environmental health."[60] That term ("environmental health") possesses a doubled meaning: what is at stake is not just human

---

[56] Jefferies, *After London*, 31.
[57] Kluwick, "The global deluge," 67.
[58] John Parham, "Bleak intra-actions: Dickens, turbulence, material ecology," in *Victorian Writers and the Environment: Ecocritical Perspectives*, ed. Laurence W. Mazzeno and Ronald D. Morrison (Abingdon: Routledge, 2017), 115.
[59] Parham, "Bleak intra-actions," 115.
[60] Alaimo, *Bodily Natures*, 2, 3.

health, when exposed to environmental risks, but the health of the environment itself, on whose functioning humankind depends in ways it is still only now beginning to understand fully.

This recognition – of the finitude of human understanding – links with the second of the two main points that follow from Jefferies's remarkable account. As we have seen, the reality to which the novel pays attention is not simply an effect designed to bolster a Barthian illusion that this is a "real" world, in which readers may freely immerse themselves for the sake of their own entertainment; that nonhuman world is itself a major actor or "actant" in the novel. In this sense, we might speak of *After London* as an instance of a fiction that goes beyond realism in its mindfulness of materiality. But its mindfulness of materiality also extends to the profound and prescient recognition that, precisely because things are active, independent, and agential, they may also act in ways that are simply not expected.

This brings me to the obvious objection to the kind of reading I have offered thus far. Picking through the hints and clues that the novel's narrators offer, I have argued that there are plausible and realistic explanations for the apocalypse that the novel unfolds with such "remorseless logic," as Edward Thomas put it.[61] Yet the narrators themselves confess their own uncertainty; indeed, the unnamed scholar adds that "it may be that even when they were proceeding, the causes of the changes were not understand."[62] Given that ambiguity, that ambivalence, it may be that I am investing these clues and hints with more meaning than Jefferies ever intended, and taking more interest in them than readers ever do; as W. J. Keith puts it, "[w]e are not concerned with what happened; we are only interested in the fact that something did happen."[63] Yet it is precisely this uncertainty – this narrative refusal to offer any definitive explanation (still less any definitive conclusion to Felix's quest) – that constitutes the novel's most perceptive and modern aspect, and another aspect of what might be termed its hyper-realism or, after the work of Karen Barad, its agential realism.[64] The vulnerability of an "industrial-capitalist colossus"[65] lies not just in its over-dependence on complex technologies, or its related lack of resilience when confronted by risks of its own making (like the impact of

---

[61] Edward Thomas, *Richard Jefferies: His Life and Work* (London: Hutchinson & Co, 1909), 256.
[62] Jefferies, *After London*, 14.
[63] Keith, *Richard Jefferies*, 117.
[64] Barad, *Meeting the Universe Halfway*, 32-34.
[65] Ulrich Beck, *Ecological Politics in an Age of Risk*, trans. Amos Weisz (Cambridge: Polity Press, 1995), 5.

the flood waters as they reverse the flow of the cities' drains and sewers): it lies in its very failure to identify the existence of these risks. As Jefferies's novel highlights, modern societies are "risk societies," and the form of "techno-scientific" understanding on which these societies are dependent is itself implicated in the creation of risks, risks that can neither be foreseen nor fully understood.[66] In turn, these risks become, as Ulrich Beck has put it, "a form of involuntary self-refutation of scientific rationality."[67] Thus, Jefferies's novel responds to the deeper crisis created by industrial modernity and its "dominant model of progress,"[68] a crisis that (as we are now aware) encompasses a whole range of challenges, including enhanced global warming itself.

As this rereading of *After London* underlines, Jefferies's remarkable novel offers an all-too plausible account of the way in which an agential reality might rebound on a society that has discounted the nonhuman world as passive and inert, and in so doing, dismissed and disvalued its own detritus and waste. As the proto-ecological thinking of writers like Jefferies underlines, everything is connected, and nothing can ever in any final sense be "removed" or "thrown away". Nor were Jefferies's concerns isolated. In a powerful instance of what Mark Frost calls "under-recorded early influence," *After London* spurred a whole slew of "similar novels and stories," all sharing Jefferies's "emerging doubts about the long-term sustainability of industrial civilisation."[69] These instances of ecologically-inflected Victorian apocalypse narratives culminate with M. P. Sheil's *The Purple Cloud* (1901), written at the very end of the period, and itself an important progenitor of modern clif-fi;[70] but as Frost points out, the influence of Jefferies's novel can clearly be traced in much more recent novels, such as Ballard's *The Drowned World*,[71] noted earlier. *After London* is, as Frost adds, "a novel that simply refuses to go away, and one which is gaining an ever more prominent place in studies of Victorian culture."[72]

---

[66] Ulrich Beck, *Risk Society: Towards a New Modernity*, trans. Mark Ritter (London: Sage, 1992), 19, 20.
[67] Beck, *Ecological Politics*, 9.
[68] Martin Ryle, "Cli-Fi? Literature, Ecocriticism, History," in *Climate Change and the Humanities: Historical, Philosophical and Interdisciplinary Approaches to the Contemporary Environmental Crisis*, ed. Alexander Elliott, Vanita Damodaran, and James Cullis (London: Palgrave Macmillan, 2017), 143.
[69] Frost, introduction to Jefferies, *After London*, xvi.
[70] Trexler, *Anthropocene Fictions*, 87.
[71] Frost, introduction to Jefferies, *After London*, xvi.
[72] Frost, introduction to Jefferies, *After London*, vii.

Conclusion

Climate change fiction, or cli-fi, is a now rapidly expanding genre, but it is also one with a pre-history; even before anthropogenic climate change was properly understood, novels responded to the emerging environmental crisis that industrial modernity was engendering. *After London* is perhaps one of the most important of these progenitors, a flood narrative that takes a recognisably modern, industrial society, and subjects it to the renewed and resurgent agency of the nonhuman world. New materialism helps us to understand why this novel is so important. With its emphasis on an agential, material world, and the co-constitutive entanglements that inseparable bind society to it, a new materialist approach forces us to re-evaluate scenes that might simply be dismissed as symbolic or metaphorical – the wild imaginings of, say, a pioneer of speculative fiction – and interpret them literally, realistically, as expressions of real concerns rooted in a now forgotten socio-historical context. In *After London*, as we have seen, those concerns relate to a fundamental shift in society's attitude towards its own waste. It is a shift, as Jefferies recognised, from the treatment of waste as part of a closed system, in which resources are regarded as finite, to an open system, in which it is assumed that resources are infinite, and sinks, endless; that there is no end to the amount that society can consume or discard. *After London* is a graphic and wholly plausible illustration of what may come of this kind of thinking, a kind of thinking which is still our own; waste cannot in any final sense be removed or discarded, even when we believe it may. But the most frightening and prescient aspect of the novel's heightened sense of agential realism is its recognition that, given a lively and dynamic nonhuman world, we cannot be sure of its own actions and reactions. As Richard Jefferies dramatized so powerfully in *After London*, we have entered a new phase of human history, in which risk is embedded; but risk may also exist beyond the limits of our own control or comprehension.

References

Alaimo, Stacey. *Bodily Natures: science, environment, and the material self*. Bloomington: Indiana University Press, 2010.
Barad, Karen. *Meeting the Universe Halfway: Quantum Physics and the Entanglement of Matter and Meaning*. London: Duke University Press, 2007.
Beck, Ulrich. *Risk Society: Towards a New Modernity*. Translated by Mark Ritter. London: Sage, 1992.

—. *Ecological Politics in an Age of Risk*. Translated by Amos Weisz. Cambridge: Polity Press, 1995.

Bennett, Jane. *Vibrant Matter: a political ecology of things*. London: Duke University Press, 2010.

Besant, Walter. *The Eulogy of Richard Jefferies*. New York: Longmans, Green & Co, 1888.

Flanders, Judith. *The Victorian City: Everyday Life in Dickens' London*. London: Atlantic, 2012.

Fleming, James Rodger. "James Croll in Context: The Encounter between Climate Dynamics and Geology in the Second Half of the Nineteenth Century". *History of Meteorology* 3 (2006): 43-54.

Fowles, John. Introduction to Richard Jefferies, *After London, or Wild England*, vii-xxi. Oxford: Oxford University Press, 1980.

Frost, Mark. Introduction to Richard Jefferies, *After London; or Wild England*, vii-xlvi. Edited by Mark Frost. Edinburgh: Edinburgh University Press, 2017.

Garnett, David. Introduction to Richard Jefferies, *After London and Amaryllis at the Fair*, vii-xiii. London: J. M. Dent & Sons, 1939.

Ghosh, Amitav. *The Great Derangement: Climate Change and the Unthinkable*. London: University of Chicago Press, 2016.

Halliday, Stephen. *The Great Stink of London: Sir Joseph Bazalgette and the Cleansing of the Victorian Capital*. Stroud: Sutton Publishing, 2001.

Hulme, Mike. *Why We Disagree About Climate Change: Understanding Controversy, Inaction and Opportunity*. Cambridge: Cambridge University Press, 2009.

Jefferies, Richard. *After London; or Wild England*. Edited by Mark Frost. Edinburgh: Edinburgh University Press, 2017.

Johns-Putra, Adeline. "Climate change in literature and literary studies: from cli-fi, climate change theatre and ecopoetry to ecocriticism and climate change criticism." *WIREs Clim Change* 7 (2016): 266-282. https://onlinelibrary.wiley.com/doi/abs/10.1002/wcc.385.

Keith, W. J. *Richard Jefferies: A Critical Study*. London: Oxford University Press, 1965.

Kluwick, Ursula. "The global deluge: floods, diluvian imagery, and aquatic language in Amitav Ghosh's *The Hungry Tide* and *Gun Island*." *Green Letters – Studies in Ecocriticism; Special Issue: Waters Rising* 24, No. 1 (2020): 64-78.

Levine, Caroline. "Victorian realism". In *The Cambridge Companion to the Victorian Novel*, edited by Deidre David, 84-106. Cambridge: Cambridge University Press, 2012.

Looker, Samuel, editor. *The Notebooks of Richard Jefferies*. London: Grey Walls Press, 1948.
Luckin, Bill. *Death and Survival in Urban Britain: Disease, Pollution and Environment 1800-1950*. London: Bloomsbury Academic, 2015.
Milner, Andrew and Burgmann, J. R. "A Short Pre-History of Climate Fiction." *Extrapolation* 59, no. 1 (2018): 1-23.
Parham, John. "Bleak intra-actions: Dickens, turbulence, material ecology." In *Victorian Writers and the Environment: Ecocritical Perspectives*, edited by Laurence W. Mazzeno and Ronald D. Morrison, 114-129. Abingdon: Routledge, 2017.
Ryle, Martin. "Cli-Fi? Literature, Ecocriticism, History". In *Climate Change and the Humanities: Historical, Philosophical and Interdisciplinary Approaches to the Contemporary Environmental Crisis*, edited by Alexander Elliott, Vanita Damodaran, and James Cullis, 143-158. London: Palgrave Macmillan, 2017. https://doi.org/10.1057/978-1-137-55124-5_7
Schneider-Mayerson, Matthew. "The Influence of Climate Fiction: An Empirical Survey of Readers." *Environmental Humanities* 10, no. 2 (2018): 473-500. https://doi.org/10.1215/22011919-7156848.
Scott, Heidi C. M. *Chaos and Cosmos: Literary Roots of Modern Ecology in the British Nineteenth Century*. University Park, PA: Pennsylvania State University Press, 2014.
Suvin, Darko. *Victorian Science Fiction in the UK: The Discourses of Knowledge and of Power*. Boston: G. K. Hall, 1983.
Tait, Adrian. "Environmental Crisis, Cli-fi, and the Fate of Humankind in Richard Jefferies' *After London* and Robert Harris' *The Second Sleep*." *Exchanges: The Interdisciplinary Research Journal, Special Issue: Climate Fiction* 8, no. 2 (2021): 69-83. https://doi.org/10.31273/eirj.v8i2.
Thomas, Edward. *Richard Jefferies: His Life and Work*. London: Hutchinson & Co, 1909.
Trexler, Adam. *Anthropocene Fictions: The Novel in a Time of Climate Change*. London: University of Virginia Press, 2015.

CHAPTER THREE

AN ECOCRITICAL ANALYSIS OF THE NATURAL LIFE IN URSULA K. LE GUIN'S *THE DISPOSSESSED* (2002)

SEHER ÖZSERT[1]

Introduction

Several inquisitive explorations and assumptions about the existence of life on other planets have been made so far. Thanks to technological advances, the search for traces of life in outer space has become a dream for mankind. This dream of an alternative life in outer space has also been appealing subject matter of both science-fiction narratives and common parlance. Ursula K. Le Guin depicts this fantasy in her book entitled *The Dispossessed* (2002). The narrative takes place on two different planets in a time span of two or three centuries after the present day. However, the story is far from being a standard high-tech outer-space adventure. In fact, the portrayal of these two planets' societies is very complicated. Le Guin foresees "An Ambiguous Utopia", the subtitle of the novel, which refers to the many possibilities between personal freedom and social/ecological duties. The reality accompanying this ambiguity in Le Guin's narration gives power to the social and scientific possibilities, as Judah Bierman points out: "Knowledge has a gift and a moral obligation to power, ambiguous but real, perhaps essential to its survival."[2] The narrative provides the apocalyptic history of the planet Earth as well, which has been completely abandoned after natural disasters. The primary reason for leaving the old planet behind in search of a new one is explicitly stated as

---

[1] Assist. Prof. Dr., Nişantaşı University, Dept. of English Language and Literature, Istanbul, Turkey. seher.ozsert@nisantasi.edu.tr
[2] Judah Bierman. "Ambiguity in Utopia: 'The Dispossessed'." *Science Fiction Studies*, No. 3 (1975), 255.

humanity's awareness of the irremediable destruction of Earth and the eventual search for an alternative habitat. The cruel idea of exploiting natural resources to the bones and the destruction of nature in an unrepairable manner are the main motives for settling on a new planet. One of the locations in Le Guin's narration, Anarres, presents a cautious human life where there is natural scarcity and the ground is infertile, like a desert. On the other hand, Urras, as the complete opposite of Anarres, presents a life of abundance with fertile land and diverse inhabitants.

Science fiction narratives stimulate readers to question the world they live in and present alternative views. Utopian novels, therefore, "focus on the continuing presence of difference and imperfection within utopian society itself and thus render more recognizable and dynamic alternatives."[3] As Tom Moylan notes, utopian fiction does not attempt to create a fixed future in readers' minds. On the contrary, it "serves to stimulate in its readers a desire for a better life and to motivate that desire towards action by conveying a sense that the world is not fixed once and for all."[4] In particular, from an ecocritical perspective, these narratives push the limits of other possibilities when the connection between human and nature is reshaped, because they focus more "on a quest for what has been repressed or denied."[5] According to Patrick D. Murphy, science fiction owes a huge thanks to these imaginations: "Rather than providing the alibi of a fantasy— in the sense of an escape from real-world problems—extrapolation emphasizes that the present and the future are interconnected."[6] Today's actions will eventually have repercussions on the future of humankind and nature, and, therefore, careful precautions must be taken to avoid a disaster. This is also reflected and discussed in science fiction narratives. Colin Milburn maintains that even the most absurd and unrealistic science fiction stories could be useful sources of "inspiration" and "influence" for people by presenting modifiable futures.[7] Such a warning could be a great opportunity: it might provide a second chance to create a better future.

In a similar vein, Rob Latham illustrates the role of science fiction from a distinctive perspective, which situates human beings as invaders of natural worlds. He asserts that "rather than the victims of biotic invasion,

---

[3] Tom Moylan. *Demand the Impossible: Science Fiction and the Utopian Imagination* (New York: Methuen, 1986), 11.
[4] Ibid., 35.
[5] Ibid., 34.
[6] Patrick D. Murphy. *Ecocritical Explorations in Literary and Cultural Studies: Fences, Boundaries, and Fields* (New York: Lexington Books, 2009), 89.
[7] Colin Milburn. "Modifiable Futures: Science Fiction at the Bench." *Isis*, No. 3 (2010): 566.

earth people are the invaders; and rather than seeding a host of trees, they lay waste to a vast forest on the planet."[8] He claims that there is not much of a difference between western colonialism and the exploration of space for inhabitation by human societies. As will be analysed here, this assessment is fully reflected in Le Guin's narration.

Tim Libretti places emphasis on the criteria of assessing and determining humanity's role especially in "the organization of work; the relationship between self, society, and nature; and the distribution of both social responsibilities and resources."[9] He states that there must be perfect unity among these three freely to sustain individual happiness and benefits for both society and nature. Peter G. Stillman maintains a similar view, arguing that human beings must work with time and nature, not against them; thus, according to him, the role of humans in nature is not controlling it but rather accompanying it.[10] He highlights the significance of the inspirational power of nature in science fiction. However, in *The Dispossessed*, it can only inspire the people who are open to it, such as Takver and Shevek.[11]

Daniel P. Jaeckle takes Le Guin's novel in terms of the prevalent idea of anarchy, questioning the ethical dilemmas that the characters face "for the explicit purpose of challenging the balance between freedom and responsibility."[12] In the scarcity of natural resources, Shevek's task of distributing food reveals the difficulty in reconciling individual choices and communal survival. Mario Klarer reads Le Guin as a female writer of science fiction, focusing on her gender-specific approach. Thus, according to Klarer, she depicts "the fertile Urras" and "the barren soil and the adverse nature of [Anarres]" through a female voice, defending simultaneity.[13] He further compares the sexual intercourse of lovers to the interchange between the two planets. This chapter will offer an ecocritical analysis of natural life

---

[8] Rob Latham. "Biotic Invasions: Ecological Imperialism in New Wave Science Fiction." *The Yearbook of English Studies*, No. 2 (2007): 114.

[9] Tim Libretti. "Dispossession and Disalienation: The Fulfillment of Life in Ursula LeGuin's *The Dispossessed*." *Contemporary Justice Review: Issues in Criminal, Social, and Restorative Justice*, No. 3 (2007): 305.

[10] Peter G. Stillman. "*The Dispossessed* as Ecological Political Theory" in *The New Utopian Politics of Ursula K. Le Guin's The Dispossessed*, ed. Laurence Davis and Peter Stillman (New York: Lexington Books, 2005), 64.

[11] Ibid., 56.

[12] Daniel P. Jaeckle. "Embodied Anarchy in Ursula K. Le Guin's *The Dispossessed*." *Utopian Studies*, No. 1 (2009): 84.

[13] Mario Klarer. "Gender and the 'Simultaneity Principle': Ursula Le Guin's 'The Dispossessed'." *Mosaic: A Journal for the Interdisciplinary Study of Literature*, No. 2 (1992): 108.

and human interaction with nature in Le Guin's dystopian novel *The Dispossessed* in light of the aforementioned ecocritical approaches.

Representation of Nature in *The Dispossessed*

Nature in *The Dispossessed* is distinctively portrayed by two contrasting planets through the personal observations of a brilliant physicist seeking to combine the principles of sequency and simultaneity in the so-called General Temporal Theory. This protagonist, Shevek, wants to prove the unity of two possible things: that time layers move in a linear way and that all times exist at the same time. However, people can freely move between those extremities. Through this practice, it would be possible to communicate instantly across the whole universe. Shevek's search correlates to humanity's desire to find traces of life on other celestial bodies in outer space. When the protagonist and his friends were children, they would look at the moon and discuss the idea that all intelligent beings tend to imagine the other planet they observe as their own moon:

> "I never thought before," said Tirin unruffled, "of the fact that there are people sitting on a hill, up there, on Urras, looking at Anarres, at us, and saying, 'Look, there's the Moon.' Our earth is their Moon; our Moon is their earth."
> "Where, then, is Truth?" declaimed Bedap, and yawned.
> "In the hill one happens to be sitting on," said Tirin.[14]

On both of these planets, people wonder about the possibility of life on the other. In fact, life on Anarres began almost two hundred years previously when a group of people, the Odonians, rebelled against the tyranny of the Urrasti and decided to live freely on another planet. Urras is strictly governed by a capitalist system and everything is based on profit and possession. On the other hand, Anarres has a communist system with no individual property but complete freedom without any rules and restrictions. Even though questions of politics are also clearly evident, this chapter will focus more heavily on the depiction of ecological aspects in the novel.

Both of the planets are far from perfect, which emphasizes the idea in Le Guin's narration that there is no ideal world. Although Urras possesses natural wealth and many resources, it is home to a patriarchal society. The subjugation of women, which is described as a "waste" by Shevek, is evident. The land of Anarres, however, is an infertile plain, similar to a

---

[14] Ursula. K. Le Guin. *The Dispossessed* (London: HarperCollins, 2002), 52-53.

desert. On the other hand, people live equally and freely on this planet. As Shevek's lover Takver suggests while looking at their moon, Urras, "we know that it's a planet just like this one, only with a better climate and worse people."[15] She also touches on some other points of politics including wars, the proletarian class, and inequality in society. While one class enjoys the richness of Urras, other parts of society starve to death and suffer from sickness coming with old age. Therefore, Takver asks, "when we know all that, why does it still look so happy—as if life there must be so happy?"[16] Shevek's response to that question is actually the experience that he acknowledges later: "If you can see a thing whole," he says, "it seems that it's always beautiful. Planets, lives [...] But close up, a world's all dirt and rocks. And day to day, life's a hard job, you get tired, you lose the pattern. You need distance, interval. The way to see how beautiful the earth is, is to see it as the moon. The way to see how beautiful life is, is from the vantage point of death."[17] This particular assessment of life always differs according to the vantage point. Le Guin intentionally uses these juxtapositions to inspire a more holistic worldview; as M. Teresa Tavormina states, her narration emphasizes "the interdependence of things, the necessary connections between opposites, the organic and analogical wholeness of the world. The balanced, patterned, harmonious foundation of the universe is part of the same vision."[18] According to the General Temporal Theory, this multiplicity explains the appreciation of uniqueness within splendid integrity.

Even though Takver prefers to stay away from Urras, as she is afraid of the emptiness of its beauty, she still wants her planet to remain as its moon. However, Shevek does not feel the same. After his reunion with Takver after four years of separation, Shevek dives into deeper thoughts about time, the environment, and the room in which they lie together: "Outside the locked room is the landscape of time, in which the spirit may, with luck and courage, construct the fragile, makeshift, improbable roads and cities of fidelity: a landscape inhabitable by human beings."[19] The locked room and the landscape of time symbolize the imprisonment of individual and intellectual freedom, as Chris Pak argues: "The idea that an act becomes human only when it occurs in the landscape of time, in both

---

[15] Ibid., 248.
[16] Ibid., 248.
[17] Ibid., 248-49.
[18] M. Teresa Tavormina. "Physics as Metaphor: The General Temporal Theory in 'The Dispossessed'." *Mosaic: A Journal for the Interdisciplinary Study of Literature*, No. 3/4 (1980): 57.
[19] Le Guin, *The Dispossessed*, 437.

the past and the future, embeds the individual into their immediate environment and emphasises the dynamism that undergirds physics and ecology as the basis for ethics and culture." [20] There is a perpetual connection between the past and the future within temporal landscapes, and "Terraforming is both physical and social, involving a superimposition of physical and intellectual landscapes." [21] Shevek decides to leave this uninhabitable planet, together with many others, with the hopes of a better future for his people. While departing from his own planet to go to Urras, he has a vision over the land of Anarres from the spaceship:

> It was the desert seen from the mountains above Grand Valley. [...] The edge of the plain flashed with the brightness of light on water, light across a distant sea. There was no water in those deserts [...] The stone plain was no longer plane but hollow, like a huge bowl full of sunlight. As he watched in wonder it grew shallower, spilling out its light. All at once a line broke across it, abstract, geometric, the perfect section of a circle. Beyond that arc was blackness. This blackness reversed the whole picture, made it negative. The real, the stone part of it was no longer concave and full of light but convex, reflecting, rejecting light. It was not a plain or a bowl but a sphere, a ball of white stone falling down in blackness, falling away. It was his world.[22]

The land together with the natural life on Anarres is infertile and colourless, except for one tiny fertile piece. It is a strip along the sea with long beaches in the southwest, allowing for fishing in this district, or farming by the seaside. On the other hand, the inner part of the land is arid and unsuitable for humans or any other living beings: "Inland and westward clear across the vast plains of Southwest the land was uninhabited except for a few isolated mining towns. It was the region called the Dust."[23] It is also noted that this desert has not been like this since the beginning of time. There is a reference to a past in which it was once home to a forest, which was later destroyed:

> In the previous geological era the Dust had been an immense forest of holums, the ubiquitous, dominant plant genus of Anarres. The current climate was hotter and drier. Millennia of drought had killed

---

[20] Chris Pak. *Terraforming: Ecopolitical Transformations and Environmentalism in Science Fiction* (Liverpool: Liverpool University Press, 2016), 133.
[21] Ibid., 134.
[22] Le Guin, *The Dispossessed*, 7-8.
[23] Ibid., 59.

the trees and dried the soil to a fine grey dust that now rose up on every wind, forming hills as pure of line and barren as any sand dune. The Anarresti hoped to restore the fertility of that restless earth by replanting the forest.[24]

They suffer in the Dust, especially while they are working on assignment at a post. They stink, and there is no water for bathing when they are away from the sea. Shevek and his friends complain about the sanitation and the bad smell, but there is nothing they can do: "But this camp was fifteen kilos from the beaches of the Temae, and there was only dust to swim in."[25] Shevek works with a tree planting crew in the Dust. They are trying to bring in tiny trees with trucks from the Green Mountains, which have comparatively more rain, and they plant those trees in the Dust with the hope of greening other parts of their bare planet.

Analogous to the scarcity of trees, there are not many animals on Anarres. Animal life is equally scarce, except for the fish in the sea. Thus, the land is uninhabitable for animals as well as humans and plants: "The silence, the utter silence of Anarres: he thought of it at night. No birds sang there. There were no voices there but human voices. Silence, and the barren lands."[26] In all aspects, it is a planet lacking in ecological terms as observed by the new inhabitants trying to settle down after observing the unique features of the land. For instance, there was once a green area between the mountains and the sea, which they named the Eden of Anarres as their first arrival point. Nonetheless, this place also turned out to be a desert later on: "But the Eden of Anarres proved to be dry, cold, and windy, and the rest of the planet was worse. Life there had not evolved higher than fish and flowerless plants. The air was thin, like the air of Urras at a very high altitude. The sun burned, the wind froze, the dust choked."[27] Even the Garden of Eden has turned into a desert here, signifying that the whole of the planet has ended up like a hell, not a paradise to dwell in.

In contrast to the barrenness of the land and the lifeless environment on Anarres, the people are considerate toward each other and also toward nature. Shevek's lover Takver is a perfect example with her care for nature and living things:

> Her concern with landscapes and living creatures was passionate. This concern, feebly called, "the love of nature" seemed to Shevek

---

[24] Ibid., 59.
[25] Ibid., 59.
[26] Ibid., 100.
[27] Ibid., 121.

to be something much broader than love. There are souls, he thought, whose umbilicus has never been cut. They never got weaned from the universe. They do not understand death as an enemy; they look forward to rotting and turning into humus. It was strange to see Takver take a leaf into her hand, or even a rock. She became an extension of it, it of her.[28]

Takver not only loves nature but also feels herself to be a part of that unity. This is consistent with the interconnectedness of all existing bodies in the universe from an ecocritical perspective. She has studied biology and introduces herself as "a fish geneticist" due to her job involving researching new techniques to improve edible fish stocks in the three oceans of Anarres. Therefore, her attachment to natural life is a passion. She works for the improvement of both human and animal futures. Her hopes and her dedicated research are not futile in essence, because it is certain that "the three oceans of Anarres were as full of animal life as the land was empty of it."[29] The reason for this is the complex but balanced life cycles of sea animals, which evokes Takver's enthusiasm for marine biology. This richness is not seen outside the oceans, because "on land, the plants got on well enough, in their sparse and spiny fashion, but those animals that had tried air-breathing had mostly given up the project as the planet's climate entered a millennial era of dust and dryness. Bacteria survived, many of them lithophagous, and a few hundred species of worm and crustacean."[30] In this bare land with its scarcity of natural organisms, survival is a challenging task. Still, these extreme conditions of scarcity also have positive effects on human beings as they have nothing but each other, as J. Jesse Ramirez notes: "Scarcity is the material condition under which the Anarresti suffer together, struggle together, and triumph together; it reminds them constantly of their world's fragility and motivates their collective labors to overcome it."[31] They learn the ways of survival and value the little things they do have in this harsh environment. On one occasion, Takver admires the web of natural marine life while watching fish in a water tank:

> This fish eats that fish eats small fry eat ciliates eat bacteria and round you go. On land, there's only three phyla, all nonchordates—if you don't count man. It's a queer situation, biologically speaking.

---

[28] Ibid., 242.
[29] Ibid., 242.
[30] Ibid., 243.
[31] J. Jesse Ramirez. "From Anti-Abundance to Anti-Anti-Abundance: Scarcity, Abundance, and Utopia in Two Science Fiction Writers." *RCC Perspectives,* No. 2 (2015): 87.

> We Anarresti are unnaturally isolated. On the Old World there are eighteen phyla of land animal; there are classes, like the insects, that have so many species they've never been able to count them, and some of those species have populations of billions. Think of it: everywhere you looked animals, other creatures, sharing the earth and air with you. You'd feel so much more a part.[32]

As Stillman asserts, Takver and Shevek are hope-giving characters for the future of Anarres, because they make a connection with nature and they feel themselves as a part of the land. This supports the arguments of both Milburn and Stillman that nature has an inspirational effect on people, which leads them to make positive changes for their future lives.

After the scarcity of resources and the drought on Anarres, Urras seems like a dream world for Shevek. When he first arrives there from his own dull land, it seems that nature, colours, architecture, and even human beings are all living in harmony on this planet. From his window, Shevek sees a grove of trees on a broad valley and this green vision fades into blue in the hills in the distance. Under the pale grey of the sky, the perfect combination of green and blue charms Shevek. He is deeply impressed by the beauty of the view:

> It was the most beautiful view Shevek had ever seen. The tenderness and vitality of the colors, the mixture of rectilinear human design and powerful, proliferate natural contours, the variety and harmony of the elements, gave an impression of complex wholeness such as he had never seen, except, perhaps, foreshadowed on a small scale in certain serene and thoughtful human faces.[33]

Every scene that he observes on Urras reminds him of the lack on Anarres. He remembers plains that are "barren, arid, and inchoate" and farming land that is "like a crude sketch in yellow chalk compared with this fulfilled magnificence of life, rich sense of history and of seasons to come, inexhaustible."[34] Considering the landscape, Shevek looks at the scenery and thinks to himself that "this is what a world is supposed to look like."[35] When Shevek is on Urras, he compares his past life in drought-stricken lands to the life full of water everywhere on this planet. While they could not take a shower on Anarres in the Dust, he feels like he is in a water paradise on Urras: "The deployment of water was wonderful [...] the

---

[32] Le Guin, *The Dispossessed*, 243-44.
[33] Ibid., 83-84.
[34] Ibid., 84.
[35] Ibid., 84.

bathtub must hold sixty liters, and the stool used at least five liters in flushing. This was really not surprising. The surface of Urras was five-sixths water. Even its deserts were deserts of ice, at the poles. No need to economize; no drought..."[36] Shevek further observes the richness of the natural world when he goes into the countryside. He cannot believe his eyes when he sees so many trees in one area compared to his own dusty homeland with almost nothing in it:

> What was the thicker darkness that flowed along endlessly by the road? Trees? Could they have been driving, ever since they left the city, among trees? The lotic word came into his mind: "forest." They would not come out suddenly into the desert. The trees went on and on, on the next hillside and the next and the next standing in the sweet chill of the fog, endless, a forest all over the world, a still striving interplay of lives, a dark movement of leaves in the night.[37]

The beauty of this land full of trees is a marvellous experience for Shevek; he watches the animals in the forest and is shocked by their abundance. He looks at a donkey, but he cannot identify it, as they do not have such large animals on Anarres. The deeper he goes into the countryside, the more fascinating it is for him. Through his subtle observations of the land and nature, he realizes that the tales he heard about this planet as a child were not actually true. It is, in fact, a far better place:

> He saw the farmlands, lakes, and hills of Avan Province, the heartland of A-Io, and on the northern skyline the peaks of the Meitei Range, white, gigantic. The beauty of the land and the well-being of its people remained a perpetual marvel to him. The guides were right: the Urrasti knew how to use their world.[38]

Contrary to the natural beauty and abundance of resources on Urras, however, the people do not enjoy intimate interactions with the natural world. Shevek observes people like robots with no faces among the buildings in an industrious city: "The towers of the city went up into mist, great ladders of blurred light. Trains passed overhead, bright shrieking streaks. Massive walls of stone and glass fronted the streets above the race of cars and trolleys. Stone, steel, glass, electric light. No faces."[39] From the first moment of Shevek's arrival to Urras, he begins to compare not only the

---

[36] Ibid., 82.
[37] Ibid., 28.
[38] Ibid., 106.
[39] Ibid., 27.

environment but also the lifestyles of the people, observing the inconsiderate attitudes of people towards natural resources. For instance, he is shocked to learn that the Urrasti burn papers and that their clothes for sleeping are used only once. The people on Urras value everything in terms of money and do not care whether it is harmful to the environment or not. Dr. Kimoe feels obliged to give an explanation upon Shevek's astonishment: "Oh, those are cheap pajamas, service issue—wear them and throw them away, it costs less than cleaning."[40] Shevek does not understand why they use different clothes while sleeping and why they throw everything into the trash so thoughtlessly. On Anarres, every product is valuable; people know that the production of paper and books requires much human labour and that the process causes many trees to be cut down. On Anarres, they are very mindful while using such resources. They write in very small letters on the available paper, using nearly every possible empty space. However, the people of Urras do not understand the value of such materials as they enjoy lives of plenty. Stillman argues that the perfect usage of materials in the scarcity of Anarres, in contradiction to the capitalists' extravagant use of them on Urras, is an example of anarchists' true ecological management: "Anarres shows that anarchy is ecologically sound because anarchists can use scarce resources efficiently and sustainably over generations, even when debilitating drought occurs."[41] Nonetheless, he also states that anarchy is not the only solution for the protection of natural resources; there are some other ways, as regulated by the government on Urras.

Pak evaluates the extravagant life on Urras in light of Darko Suvin's analysis of Frederick Jameson's interpretation of scarcity as "a reaction to the polluted American abundance and a realistic diagnosis of a better model of life."[42] From Ursula K. Heise's perspective, Pak further analyses that these environmental discourses of abundance and scarcity serve "as conceptual tools whose usefulness in shaping contemporary societies' relation to their environments needs to be constantly re-examined."[43] The Urrasti people have no management system for the usage of natural resources as seen in present-day societies. Being inconsiderate citizens, they are only controlled by laws restricting them in the use of certain things. For example, they are not allowed to own cars freely for fear of draining natural resources. They mostly hire cars, but hiring them is also very expensive, and privately purchased cars are heavily taxed. Therefore, very few people can afford to own cars. Through these regulations, people are kept under control

---

[40] Ibid., 16.
[41] Ibid., 56.
[42] Quoted in Pak, *Terraforming*, 128.
[43] Quoted in Pak, *Terraforming*, 129.

to avoid the contamination of the environment with waste. It is said on Urras that if they did not apply ecological control, the planet would not last for centuries. They have taken lessons from the excessive usage of the previous millennium, which resulted in the shortage of some metals. Now they import those metals from Anarres and they try to keep others strictly monitored; this, in fact, means the exploitation of another planet. "In fact, the Free World of Anarres was a mining colony of Urras."[44]

Shevek's life on Urras is admirable at first, full of routines, teaching at the university, travelling, learning new things, and adapting to the current atmosphere, both at home and outside. Nevertheless, he feels something missing in this wealth, something he is unable to name at first: "There was something lacking – in him, he thought, not in the place. He was not up to it. He was not strong enough to take what was so generously offered. He felt himself dry and arid, like a desert plant, in this beautiful oasis."[45] He had interrupted his life on Anarres; he had moved away from there, and now "the waters of life welled all around him, and yet he could not drink."[46]

After spending some time on Urras, Shevek understands that he cannot realize his ideals among these people who only consider the advantages they can gain as a result of any enterprise. The initial admiration of his first days disappears as he perceives the hopelessness of his struggle due to the absolute profit-seeking nature of the people on the planet; he comes to see that Urras is like a hell in all its beauty. He confesses his thoughts to Keng, an ambassador of Terra, while asking for help to get out of this terrible place:

> There is nothing, nothing on Urras that we Anarresti need! We left with empty hands, a hundred and seventy years ago, and we were right. We took nothing. Because there is nothing here but States and their weapons, the rich and their lies, and the poor and their misery [...] It is a box—Urras is a box, a package, with all the beautiful wrapping of blue sky and meadows and forests and great cities. And you open the box, and what is inside it? A black cellar full of dust, and a dead man. A man whose hand was shot off because he held it out to others. I have been in Hell at last. Desar was right; it is Urras; Hell is Urras.[47]

Shevek observes Urras as hell. Keng, on the other hand, envisions it as heaven. The reason for these different perceptions is obvious, as Jaeckle

---

[44] Le Guin, *The Dispossessed*, 119.
[45] Ibid., 175.
[46] Ibid., 175.
[47] Ibid., 452-53.

also notes: "The planet [Keng] comes from is a future version of Earth trying desperately to recover from near annihilation due to the pollution and war. Compared to the gray heat of Terra, Urras is a most beautiful planet."[48] Keng feels that this planet is at least alive compared to the dead one that she left behind: "I know it is full of evils, full of human injustice, greed, folly, waste. But it is also full of good, of beauty, vitality, achievement. It is what a world should be! It is alive, tremendously alive – alive, despite all the evils, with hope."[49] Nevertheless, Shevek has no hopes on Urras, but he does have hopes for the future of his people and other people living on other planets with the help of his theory. He does not want to share his theory with the selfish and spoiled people on Urras; he only wants to share it with other societies as an act of kindness. In contrast to Shevek, Ambassador Keng has no hope for the future. However, she admits that she is capable of being satisfied with the present conditions on Urras. Through this Terran ambassador, there is a reference to Earth as a planet destroyed centuries ago by the thoughtless people on it. In their conversation, Keng reveals to Shevek that their world was once as beautiful and fertile as Urras, but they turned their paradise into a hell:

> My world, my Earth is a ruin. A planet spoiled by the human species. We multiplied and gobbled and fought until there was nothing left, and then we died. We controlled neither appetite nor violence; we did not adapt. We destroyed ourselves. But we destroyed the world first. There are no forests left on my Earth. The air is grey, the sky is grey, it is always hot. It is habitable, it is still habitable, but not as this world is. This is a living world, a harmony. Mine is a discord. You Odonians chose a desert; we Terrans made a desert.[50]

Keng does not have any hope for the future of her planet, so she recognizes Urras as a paradise. Shevek disagrees with her; he believes that Keng's world might have been destroyed and people might live in misery, but still, this does not mean the same for Anarres: "You say the past is gone, the future is not real, there is no change, no hope. You think Anarres is a future that cannot be reached, as your past cannot be changed. So there is nothing but the present, this Urras, the rich, real, stable present, the moment now. And you think that is something which can be possessed!"[51] Shevek does not approve of extravagant *carpe diem* moments, and he believes that one must consider the past and the future together, acting accordingly,

---

[48] Jaeckle. "Embodied Anarchy in Ursula K. Le Guin's *The Dispossessed*," 89.
[49] Le Guin, *The Dispossessed*, 347.
[50] Ibid., 454.
[51] Ibid., 456.

because "they are real: only their reality makes the present real."[52]

Stillman criticizes this idea of over-emphasis on place, which determines the success or failure of the society. He challenges Keng's view that if there is plenty in a place, it makes it a paradise, but if there is scarcity, then it is a hell. He highlights the significance of the point that neither geography nor material wealth determines a society's success and that people do not need to be connected to any particular place, as on Anarres.[53] The people on Anarres reflect this clearly: they can easily change their places, positions, and friends. They adapt to new environments easily, because they try to assume that "home is a place where you have never been."[54] In this sense, the novel also emphasizes the interconnectedness of all nations and places through Shevek's struggle to connect them by tearing down walls as described at the very beginning. Stillman observes that Le Guin is not rejecting the variety of habitats and experiences within a wide range of geographical differences; "rather, she is rejecting the walls or boundaries that separate people and seeing others as an opponent or enemy. Just as Shevek sees the connectedness of time, he wants to tear down walls that break the connectedness of space."[55]

Conclusion

Through the juxtaposition of fictional planets at two extremes, Le Guin warns that if wise management of natural resources is not maintained, the end of our planet is inevitable. As Le Guin's mouthpiece, Ambassador Keng explains how they manage to survive: "Well, we had saved what could be saved, and made a kind of life in the ruins, on Terra, in the only way it could be done: by total centralization. Total control over the use of every acre of land, every scrap of metal, every ounce of fuel."[56] The only way for them to continue their lives on Terra is, as she desperately acknowledges, the total centralization of authority in every field. This is something neither Shevek nor other Odonians would agree on. Murphy argues that the destruction of the biodiversity on the planet also restricts human freedom, in addition to self-destruction, and in Le Guin's fiction "it is not environmentalists who would deny freedom to other human beings but that the unbridled consumption of the present day will necessitate draconian

---

[52] Ibid., 456.
[53] Stillman, "*The Dispossessed* as Ecological Political Theory," 66.
[54] Le Guin, *The Dispossessed*, 70.
[55] Stillman, "*The Dispossessed* as Ecological Political Theory," 66.
[56] Le Guin, *The Dispossessed*, 455.

measures tomorrow."[57] He also recalls Le Guin's warning to humanity that a promising future depends on people's comprehension of the sensitivity implemented in the ideals of the anarchists on Anarres, not the lavishness on Urras. By the end of the book, people start to question the current systems. They try to understand Shevek's arguments and his description of Anarres. In response to a question, Shevek makes a successful comparison of the worlds: "No. It is not wonderful. It is an ugly world. Not like this one. Anarres is all dusty and dry hills. All meager, all dry. And the people aren't beautiful [...] The towns are very small and dull, they are dreary. No palaces. Life is dull, and hard work. You can't always have what you want, or even what you need, because there isn't enough."[58] Shevek's utterance reveals the awareness that he has acquired after lifting the veil of the splendid material world and facing the calamitous reality behind that.

Without a doubt, Anarres is a land of scarcity. The land and the appearances of the people living there are not depicted as being beautiful as on Urras. The only nice thing on Anarres as described by Shevek is the happy faces of its citizens, the most important thing of all. He describes Urras as a planet of plenty in all ways, but the Urrasti people do not have happy faces, as they are not free like the Anarresti. These comparisons of beauty reveal the idea that appearances may be deceiving, and one must analyse the situation more deeply to determine whether the land is a dream or a prison. Shevek interprets the planet of Urras as a prison, where the prisoners wear jewellery. Richness is meaningless without freedom according to him. Therefore, he decides to escape from Urras and return to his free life on the ugly land of Anarres. He keeps in mind the lesson written on Odo's gravestone: "To be whole is to be part; true voyage is return."[59]

In Le Guin's *The Dispossessed*, these two distinctive planets reveal a great deal of criticism of the exploitation of natural resources and its dire consequences for humankind. While the untouched natural beauty on Urras is described with admiration, the infertile and dull land of Anarres is depicted as a wasteland. However, this juxtaposition is deconstructed towards the end. Shevek's dissatisfaction and disgust with the state of Urrasti citizens leads to his awakening and eventual departure from that planet. Thus, the novel serves as an ecological warning about the depletion of natural resources and the destruction of nature. The citizens of Urras have become a consumerist society, seeking only material possessions. In spite of the restrictive governments and other dystopic elements in the novel, Le Guin still offers readers a glimpse of hope in the end: it all depends on the

---

[57] Murphy, *Ecocritical Explorations in Literary and Cultural Studies,* 95.
[58] Le Guin, *The Dispossessed,* 299.
[59] Ibid., 108-9.

will of humanity to act to save the world we live in. Shevek and other such considerate individuals can even save the bare land of Anarres for the sake of humankind and other living beings. As Stillman notes, a "hopeful future lies in the knowledge that Anarres exists as an ongoing and feasible attempt to create and live a better world, a continuing process or seeking of freedom and mutuality."[60] Consequently, the future lies in the hands of conscious individuals and in their attempts at working towards a better future for all of humanity. Le Guin perfectly presents this message as encouragement to the reader to make radical decisions that will influence the next generations by changing the course of our ecologic history. The theory of simultaneity within the narration, as an exceptionally developed version of Einstein's relativity, also supports the idea that everything in this universe has a tendency of leaking into everyone's lives. This reveals the significance of today's actions with unpredictable reflections as time, space, and all living bodies are intimately interconnected pieces, acting harmoniously with each other.

References

Bierman, Judah. "Ambiguity in Utopia: 'The Dispossessed'." *Science Fiction Studies*, No. 3 (1975): 249-55.
http://www.jstor.org/stable/4238976.
Jaeckle, Daniel P. "Embodied Anarchy in Ursula K. Le Guin's *The Dispossessed*." *Utopian Studies*, No. 1 (2009): 75-95.
http://www.jstor.org/stable/20719930.
Klarer, Mario. "Gender and the 'Simultaneity Principle': Ursula Le Guin's 'The Dispossessed'." *Mosaic: A Journal for the Interdisciplinary Study of Literature,* No. 2. (1992): 107-21.
http://www.jstor.org/stable/24780621.
Latham, Rob. "Biotic Invasions: Ecological Imperialism in New Wave Science Fiction" *The Yearbook of English Studies*, No. 2 (2007): 3-19.
https://doi.org/10.2307/20479304
Le Guin, Ursula. K. 2002. *The Dispossessed*. London: Harpercollins.
Libretti, Tim. "Dispossession and Disalienation: The Fulfillment of Life in Ursula LeGuin's *The Dispossessed" Contemporary Justice Review: Issues in Criminal, Social, and Restorative Justice*, No. 3 (2007): 305-20. https://doi.org/10.1080/1028258042000266022

---

[60] Stillman, "*The Dispossessed* as Ecological Political Theory," 68.

Milburn, Colin. "Modifiable Futures: Science Fiction at the Bench" *Isis*, No. 3 (2010): 560-69. https://doi.org/10.1086/655793

Moylan, Tom. 1986. *Demand the Impossible: Science Fiction and the Utopian Imagination.* New York: Methuen.

Murphy, Patrick D. 2009. *Ecocritical Explorations in Literary and Cultural Studies: Fences, Boundaries, and Fields.* New York: Lexington Books.

Nadir, Christine. "Utopian Studies, Environmental Literature, and the Legacy of an Idea: Educating Desire in Miguel Abensour and Ursula K. Le Guin." *Utopian Studies*, No. 1 (2010): 24-56. https://doi.org/10.5325/utopianstudies.21.1.0024

Pak, Chris. 2016. *Terraforming: Ecopolitical Transformations and Environmentalism in Science Fiction.* Liverpool: Liverpool University Press.

Philmus, Robert M. 2005. *Visions and Re-Visions: (Re)constructing Science Fiction.* Liverpool: Liverpool University Press.

Ramirez, J. Jesse. "From Anti-Abundance to Anti-Anti-Abundance: Scarcity, Abundance, and Utopia in Two Science Fiction Writers" *RCC Perspectives,* No. 2 (2015): 83-90. https://www.jstor.org/stable/10.2307/26241320.

Stillman, Peter G. "*The Dispossessed* as Ecological Political Theory" In *The New Utopian Politics of Ursula K. Le Guin's The Dispossessed,* edited by Laurence Davis and Peter Stillman, 55-74. (New York: Lexington Books, 2005).

Tavormina, M. Teresa. "Physics as Metaphor: The General Temporal Theory in *The Dispossessed*" *Mosaic: A Journal for the Interdisciplinary Study of Literature*, No. 3/4 (1980): 51-62. http://www.jstor.org/stable/24780261.

Williams, Donna Glee. "The Moons of Le Guin and Heinlein." *Science Fiction Studies*, No. 2 (1994): 164-72. http://www.jstor.org/stable/4240331.

CHAPTER FOUR

THE APOCALYPSE OF FREE WILL IN *THE DISPOSSESSED*

PINAR SÜT GÜNGÖR

Introduction

The argument concerning free will and determinism during the 21st century is one of the most controversial or, rather, self-contradictory ones. The discussion roughly descends from ancient philosophers to the present day. This study's main concern is to signify the literal consequence of the Anthropocene and the apocalyptic vision of climate fiction, often abbreviated as "cli-fi," in terms of free will and determinism. In a climatically changed world, "novels may present a particularly productive space in which to imagine and think through climate crisis, and especially what it means to live it."[1] In what follows, "the imaginative capacities of the novel have made it a vital site for the articulation of the Anthropocene."[2] Thus, this study represents major efforts to depict human experience, covered with deterministic repressions, in fictional futures via *The Dispossessed* by Ursula K. Le Guin. As a sophisticated starting point, it is necessary to reveal the main paradigms of free will and determinism to facilitate a closer examination. "Determinism is the thesis that the past and the laws of nature together determine a unique future, that only one future is consistent with the past and the laws of nature."[3] As Inwagen claims, in moral territories, it is asserted that the human being, as a morally

---

[1] Astrid Bracke, "Flooded Futures: The Representation of the Anthropocene in Twenty-First- Century British Flood Fictions," *Critique: Studies in Contemporary Fiction* 60, no. 3 (2019): 280. https://doi.org/10.1080/00111619.2019.1570911.
[2] Adam Trexler, *Anthropocene Fictions: The Novel in a Time of Climate Change* (Charlottesville: University of Virginia Press, 2015), 23.
[3] Peter Van Inwagen, "When is the Will Free?" *Philosophical Perspectives* 3 (August 1989): 400. http://www.jstor.org/stable/42920381.

responsible creature, has to consider necessitation and the laws of nature to be in harmony with the real world. If any alternative explanation is needed, it may be correlated with the terms "ascription of responsibility," "provisos," "laws," "morality," and naturally enough, "doctrines," and "threats." Admittedly, according to deterministic approaches, agents are responsible for what they have done. To put it another way, in these kinds of cases, the individual is directly or indirectly manipulated and is tempted to choose a morally acceptable path to be praised and approved by their social environment. "It proves as a matter of fact that we are sometimes irresistibly subject to another's will,"[4] writes Bergson to signal the concrete reality between conscious forces and free will. This average response spectrum, naturally, hinders personal choices with the concern that social approval may be lost. Therefore, it can be deduced that determinism, in analytic and philosophical contexts, may be associated with punishments, social pressures of all kinds, and dominant effects on individuals.

Free will, primarily, is incompatible with deterministic approaches. "Presupposition" about free will, which is a philosophical term, corresponds to humanistic evaluations that accept the agent as a decision mechanism that will restrict determinism moving forwards. Fischer describes this sequence as "facts about the agent's actual reasoning, decision, and action."[5] Further facts about free will empower a libertarian hypothesis that modifies social concerns and restrictions in terms of labels and classical relics. As far as this study is concerned with the so-called free will of characters in Ursula K. Le Guin's *The Dispossessed*, the crucial question at stake in this debate is whether the characters have free will in a utopian world that was founded by liberal agents to become free of any kind of hierarchical state restrictions. Collective conscious and moral accountability, as the primary causes of deterministic attitudes, hinder at least some affairs of free will in the novel, which is partly against these kinds of accounts. To see why, we need to understand the reason why someone is unable to act as a free agent and how heredity and environment manipulate their deliberations and choices. Although some critics might quarrel with this assertion, we have grounds to claim that determinism and free will are hand in hand in *The Dispossessed*. To put it another way, Ursula K. Le Guin depicts the hidden reality that in most cases, if humans are present, free will might turn into determinism like going backward in time. Somewhat surprisingly, this confused situation in

---

[4] Henri Bergson, *Time and Free Will* (New York: Dover Publications, 2016), 157.
[5] John Martin Fischer, *The Metaphysics of Free Will* (Oxford: Blackwell Publishers, 1994), 408.

the novel is not very clear since readers are satisfied with compromises provided by a free and ungoverned planet with a utopian view. With this fact in mind, heredity-environment controversy becomes the main discussion point of the study. Smart writes, "We believe that heredity, accident, and incident have a bearing on man's character and actions, and may even sometimes have a determinative one."[6] Just as we might praise or dispraise this thought, it is widely accepted in social studies. Admittedly, to provide proof for the idea that "a man's drive is determined by his genes and his environment,"[7] it is necessary, naturally, to discuss the author, plot, and setting of the novel to get a better understanding of individual and environmental drives.

Ursula K. Le Guin's Biography

Ursula Kroeber Le Guin was born in 1929 in California. She is a well-known and appreciated author of 21 novels and many other collections. The breadth and topics of her fiction earned her many important prizes, like the Nebula Award, Hugo Award, PEN/Malamud Award, and many others. Because of her family—her father was an anthropologist and her mother was a writer—she met people from all walks of life. This acquaintance, innately, enriched her fiction and diversified her themes. All in all, growing up in an intellectual family with rich learning environments shaped Le Guin's fiction. Her obvious fascination with idealistic organizations and her attempt to identify anarchism find a place in science-fiction tradition. "She imagines a society without three great enemies of freedom: the state, organized religion, and private property."[8] In the intellectual world of the late 20th century, ethnicity, gender issues, disciplinary states, consumerism, and traditional politics are topics that are outdated and harshly criticized. Being among the post-modern anarchist tradition, Le Guin literally redesigned science fiction with a radical shift from these controversial subjects. More importantly, free will, climate change, sexuality, and anarchy, which are entirely relevant to post-modern tradition, are genuinely included in her fiction. Remarkably, Le Guin updated traditional anarchist concepts with a critical view that enhanced the positions and natures of literary approaches

---

[6] Jack J. C. Smart, "Free-Will, Praise and Blame," *Mind* 70, no. 279 (1961): 292, http://www.jstor.org/stable/2251619
[7] Ibid., 305.
[8] Daniel P. Jaeckle, "Embodied Anarchy in Ursula K. Le Guin's *The Dispossessed*," *Utopian Studies* 20, no. 1 (2009): 75, http://www.jstor.org/stable/20719930

of a new age. By advocating for essential and complicated themes in this way, she detected how humans might change an idealistic world into a common one through the attainment of an invisible deep analysis.

Authentic anarchism, according to Le Guin, should reject any kind of oppression, limitation, and restriction. More urgently, it should make a deal between individual and society. Yet, by implicating hidden collective repression in *The Dispossessed*, she endeavors to point to individuals' desire for power amalgamated with autonomous will. Written in 1974, *The Dispossessed* is an ambiguous utopian work that redefines the scope and style of mainstream utopian tradition. In this novel, Le Guin introduces two planets, Urras and Anarres, to the reader as complementary. On the most basic level, these two planets, hugely different from each other, are instances of the real world and a utopia. Seemingly, the two most distinctive slices of reality in the novel are implied to further the possibility of an ungoverned, free, unlimited new world, Anarres, that provides its inhabitants large-scale opportunity with complete free will. This idealistic view, paradoxically, is damaged by unavoidable social responsibility paradigms that force characters to abandon, to some extent, free will and suppress their creativity. So implicit is this trapping that most critics argue that individual freedom surpasses any kind of governance in Anarres. Yet the basic dilemma is that in Anarres, there is no government, no gender issues, no religion, no hierarchy, and no traditional institutions that coerce people to act against their free will; however, toward the end of the novel, it becomes clear that even in that idealistic world that was founded to react against any kind of possession, any kind of ruling, collective consciousness and social responsibilities make people behave according to the climate of the place they live in. The main character Shevek, a physicist from the anarchist society of Anarres, makes unusually successful studies in his field. Nonetheless, due to the challenges of the physical environment of Anarres, he, like any other citizen of the planet, works as a common person to reap the social benefits. At this point, it is vitally important to discuss the harsh physical conditions of Anarres to get a better understanding of journeys from Anarres to Urras, which seem to be paradoxical since these planets are political enemies. Besides, it is relevant for considering whether the world of *The Dispossessed* should be categorized as utopian or dystopian.

The (Non) Existence of Free Will in *The Dispossessed*

In its very concreteness, Anarres is a rather dry and barren planet, due to features that always make it prone to droughts and scarcities.

Relatedly, species of animals and plants are few in number and are not populous enough even for daily use. The climate of this planet is covered in dust and makes difficult the sustainability of human life, which is threatened with extinction and epidemics. The hard life conditions force people to diminish the dreadful effects of their physical environment. This effort, significantly, takes the form of a kind of totalitarian control that is concealed under the name of social consciousness. That is, it may not be wrong to say that climate, as the basic point, is a determinant force for the mutual aid that is the single principle for the habitants in Anarres. The planet's only wealth is minerals, which makes possible it to trade with Urras—it is a probabilistic sample of the real world in this way. The question then becomes, why does Le Guin, in a utopian novel, suggest such a difficult and repressive environment that is incompatible with ideals? Perhaps the best answer, as implied in the narrative, is to create boundaries of social awareness that negotiate both new forms of anarchism and the needs of the group. With a few notable exceptions, it can be stated that Le Guin succeeds in drawing these lines; however, the renounced humanitarian goals of this society problematize any sense of individual autonomy, which is one of the most remarkable distinctions from Urras. Anarres, in the novel, embodies a representation of anarchy that provides opportunities in terms of individual rights, free will, class-free society, the right of choice in each condition, and a new setting that is completely different from old-world Urras via the premises of Odo, who was a revolutionary character for the citizens of Anarres. Up to this point, the planet can be evaluated as having an ideal utopian aim; however, Le Guin changes this ideal wish into real community life, which is challenged by the physical difficulties of the environment. Odonian society, with an intellectual background, seems to handle all problematic situations with determination and an organized program. "Each individual is a cell in the social organism and is responsible for performing its specialized function, like a red blood cell carrying oxygen to the rest of the body."[9] This quotation affirms that social obligations, especially in hard times like drought or famine that occurred due to bad weather conditions and heavy dust upon the planet, become more crucial than individual desires.

    According to Odonians, the followers of Odo (who is not alive during the narrative but whose doctrines play a significant role in the novel), sacrificing is inevitable and necessary when it is required. Yet Le Guin gives permission only to Shevek and his mate Takver to question whether the climactic system succeeds in reaching its own goals. Based on

---

[9] Jaeckle, "Embodied Anarchy," 80.

this line of thought, Shevek moves from being a social organism that was created in Anarres to contribute to the common interest to an individual who criticizes ethical arrangements that need to manage the balance between free will and responsibilities. He stands in a place between the past and the future and is not content with the idea that anarchism always means the best with all strengths it has. Most obviously, in the novel, Le Guin asserts that "the Odonian society was conceived as a permanent revolution, and revolution begins in the thinking mind."[10] This clarification explains what matters is how Shevek lives the epiphany. After being praised for his academic studies, Shevek decides to go Urras, where he can share or promote his studies, which are completely related to time matters in physics, by which it will be possible to be a pioneer. However, on Anarres, which is thought to be a free and non-governed planet that prioritizes individual choices and rights, the crucial studies of Shevek are not published. As an unpredictable reaction, this can be criticized in terms of the elimination of individual rights by a person or system that has no right to accept or disapprove of something related to its inhabitants. This hindrance, as a threshold matter, first makes Shevek and Takver disappointed. Insistently, Shevek accepts the proposal from scientists from Urras since he desperately wishes to complete his time studies; besides, he desires to realize himself. For reasons similar to those in his first disapproval, Shevek's individual choice is prevented once again on the condition that if he goes Urras, then no return is possible per Odonian doctrine. This coercion is exactly in contradiction with the spirit of Odo since if Shevek stays in Anarres, it means that he would need to abandon the individual freedom that enables him to study and make these studies known. Le Guin, with this dilemma, implies the predictability of determinism if there is any community or system responsible for arranging things in the framework of social obligations. More precisely, until that point, two related causal chains imply Le Guin's own criticism that human beings have merely a chance of freedom if there is no controversial situation that puts individuals against any sort of authority. Nevertheless, it is unfair to say that in Anarres, no freedom exists that will protect the free will of its inhabitants. However, it is fairly clear that even in Anarres, which was founded to guarantee freedoms that are completely separate from any obligation to authority, the edge between free will and social obligations becomes blurred. In any case, the choice between these two territories sometimes depends upon the conditions in which causal necessitation occurs and sometimes upon possibilities.

---

[10] Ursula K. Le Guin, *Mülksüzler* (İstanbul: Metis Yayınları, 2020), 284.

After deciding to discover his real identity on an alien planet, Shevek goes to Urras alone by leaving his mate and child on Anarres. The ascription of responsibility, even on this planet, causes much unsettlement and many dilemmas that hinder Shevek's desire to study. Indeed, heredity and environment as dominant forces, which are quite consistent with deterministic evaluations, appear in Shevek's mind while he is appraising the opportunities provided by the Urras government. The huge differences mostly disturb Shevek and, in terms of resources and obligations, put him in a place very far away from authentic self-evaluation. Rewards offered by Urras and implied punishments forwarded by Anarres contribute to his delusions for a while until he actually quits studying. The environment may be appreciated as an innate force, and with this comment in mind, Shevek is shattered by the hypocrisy of scientists in Urras. The physical environment, while it seems comfortable and beneficiary, again challenges his free will. Heredity, on the other hand, as coming from Odonian community, confronts ordinary obligations and free will to choose. In simplest terms, even how to dress, behave, or eat is determined by someone in Urras. This disappointment is such that lack of drive to complete his time theory, which will change the whole system of the planets, influences his value judgment in these so-called luxurious conditions. There is a rebellion in one part of the country, which aims to achieve freedom, at the time that his internal feud connects with his physical dilemmas, and Shevek decides to return to Anarres. Admittedly, this decision is not an easy one since he will be punished by Anarres doctrines and he does not know whether they will allow him to step on the planet or kill him to prevent such kinds of attempts. Blame, like praise, is embraced by Shevek and after a long time of self-blame and self-reproach, he returns. For one thing, Le Guin may offer in Anarres, in practical and ideological terms, equality, freedom, free will, inspiration, and social responsibility—concepts that are very distinctive from Urras, where rebellions are concealed by the state-controlled press; however, in Anarres, to some extent, all these nuances can change according to the cases.

Toward the end of the novel, Shevek and Takver question the previous cases that have been shaped by social obligations; until that time, they had considered them their own choices:

> No. The fact is, neither of us made up our mind. Neither of us chose. We let Sabul choose for us. Our own, internalized Sabul—

> convetion, moralism, fear of social ostracism, fear of being different, fear of being free! Well, never again. I learn slowly, but I learn.[11]

This quotation appraises the critical mind, which is the basis for Odonian doctrines. Yet, in time, some more equal Odonians, like Sabul, change the direction of this consideration toward society. By proposing that social responsibilities are more essential than individual ones or, in other words, the whole is something greater than its parts, Sabul and some others decide on behalf of society. Then the key question comes: Where is free will? Or why do these Odonian people accept all duties given to them by some mechanisms without question? The answer is clear and simple: "When a man feels himself alone against all the rest, he might well be frightened."[12] After all that happens, Shevek realizes that fear of being alone in a large crowd forces Odonian people to accept all duties against their free will. He finds that the concept of free will is among the unsettling doubts of even ungoverned societies like Anarres. Naturally enough, this epiphany picks holes in the Odonian chain of thought that is surrounded by freedom, equality, individualism, and free will. Intending to clarify this missed point, Shevek claims the following:

> We always think it, and say it, but we don't do it. We keep our initiative tucked away safe in our mind, like a room where we can come and say, 'I don't have to do anything, I make my own choices, I'm free.' And then we leave the little room in our mind, and go where PDC posts us, and stay till we're reposted.[13]

This is a quotation taken from a quarrel between Takver and Shevek. With this raw reality, which causes a collapse in Shevek's mind, he tries to persuade Takver that no free will is actually permitted in Anarres. Although Takver insists on not believing this fact, Shevek helps her to see the truth with the question he asks: "Listen, you tell me, how many people do you know who refused to accept a posting—even before famine?"[14] This question poses real suspense in the middle of the room and also in the mind of Takver. Accordingly, cooperation, freedom, equal rights, consciousness, dispossession, individual choices—all the terms, which are the basic elements of Odonian philosophy—lose their meanings with sudden enlightenment. Shevek explains his awakening with these sentences:

---

[11] Ibid., 283.
[12] Ibid., 280.
[13] Ibid., 280.
[14] Ibid., 281.

> Well, this. That we're ashamed to say we've refused a posting. That the social conscience completely dominates the individual conscience, instead of striking a balance with it. We don't cooperate—we obey. We fear being outcast, being called lazy, dysfunctional, egoizing. We fear our neighbor's opinion more than we respect our own freedom of choice. You don't believe me, Tak, but try, just try stepping over the line, just in imagination, and see how you feel.[15]

As seen in the quotation, Shevek is instrumental in bringing critical mind issues to the narrative, and he clearly proclaims the possible consequences of refusing a determined task in terms of social and individual aspects. The message retrieved from his personal opinions, combined with his past experiences, put abolishment of freedoms idea straight in this excerpt. To say the least, Shevek deduces that for practical interests, one must accept the deterministic position given to one as a moral responsibility, the rejection of which will cause a guilty conscience. This remorse might be correlated with the effect of the environment upon human beings since in the light of these cause-effect relationships, there is no room for personal perceptions that regard only individual standpoints:

> Every emergency, every labor draft even, tends to leave behind it an increment of bureaucratic machinery within PDC, and a kind of rigidity: this is the way it was done, this is the way it is done, this is the way it has to be done...[16]

None of these opinions are at all precisely clarified by Le Guin; however, she makes Shevek see the discrepancy between ideal Odonian society and the actual case in the realm of life. In this quotation, Shevek infers the true nature of change, the process in which moral determination surpasses individual rights to choose:

> A healthy society would let him exercise that optimum function freely, in the coordination of all such functions finding its adaptability and strength. That was a central idea of Odo's *Analogy*. That the Odonian society on Anarres had fallen short of the ideal did not, in his eyes, lessen his responsibility to it; just the contrary.[17]

---

[15] Ibid., 281.
[16] Ibid., 281.
[17] Ibid., 284.

This self-questioning, subsequently an inward criticism covered by possible cases that caused this failure, certifies the fact that Odonian philosophy throughout time has undergone a number of changes, and this process turns into a force detrimental to Odonian ideals. In moral contexts where rules are of the utmost importance, accepting the concrete reality of the diminishing free will of individuals is difficult. Yet Le Guin turns her utopian novel into a dystopian one in this way by letting some characters witness the restriction placed upon their minds. She clearly distinguishes between the questions of whether a utopian planet could provide endless freedom and free will for its inhabitants or, like in the real world, whether they have to obey social responsibilities even if those responsibilities conflict with their desires. This dilemma, which dominates the narrative, can be handled in a discursive and episodic way; the former has been already clarified with Shevek's statements. When it comes to the episodic manner, it should conform to events that occur in this utopian novel. Hence, Le Guin's unique style provides many episodic incidents that describe the apocalypse of something and, at the same time, point to new beginnings.

An apocalypse concerning the struggle between good and evil in the subtext emerges in a different manner in *The Dispossessed*. In classic utopian works, apocalypse is linked with evil, and it mostly appears in the form of punishment. However, Le Guin, in subtext, matches it not only to Urras but also to Anarres, the utopian one. With these apocalyptic scenarios, she aims to mark that "even the most egalitarian utopias must take into account the competition and conflict that will arise for scarce resources."[18] *The Dispossessed* exemplifies the emotional apocalypse of inhabitants due to hard work, scarce resources, emotional deprivation, lack of intimacy in individual relations, bad weather conditions, and strictly programmed lifestyles. Conditions that can be evaluated as eco-pathological hazards within this environmental canon clarify the author's fiction, which is in contradiction with utopia. Rather than being happy and peaceful, all characters in the novel are seen as agents fulfilling a moral responsibility to survive. And Shevek, as the main character, exemplifies the loneliness of a moral person at the center of sociality. Although it is stated that the characters do all the tasks with great enthusiasm, it is possible to say that the job description in their inner world is different. "The phrase 'this is my duty' does not mean much beyond 'I will be punished if I do not do this'. The idea of duty has an external meaning, but it has no inherent

---

[18] Greg Garrard, *Ekoeleştiri,* trans. Ertuğrul Genç (İstanbul: Kolektif Kitap, 2020), 136.

meaning."¹⁹ When evaluated in the context of social responsibility, Shevek's questioning and internal reckoning gain meaningful integrity in which loneliness and apocalypse of free will can be understood correctly. "'The isolation of man from nature' is portrayed with a planet where no other species other than man exists."²⁰ This isolation, which causes a kind of emotional deprivation, makes itself apparent in individuals' decisions. *The Dispossessed*, an ambivalent novel in every sense, protects the relationship between psychological well-being and facing the responsibility of decisions. Yet, when individual differences are destroyed, like in Shevek's case, it is hard to maintain psychological well-being. As he is different from many others, he strives for social benefit, which has the potential to unite the whole universe under one roof:

> It was for that idea that I came here, too. For Anarres. Since my people refuse to look outward, I thought I might make others look at us. I thought it would be better not to hold apart behind a wall, but to be a society among the others, a world among the others, giving and taking. But there I was wrong—I was absolutely wrong.²¹

Admittedly, maybe for the first time in his life, he makes a choice in accordance with his free will. But he is wrong. In Anarres, which is barren in every way, it would again be ambivalent to talk about the existence of free will. As if to recall the founding purpose of Odonian society and Anarres, Shevek speaks at a meeting:

> You see, what we're after is to remind ourselves that we didn't come to Anarres for safety, but for freedom. If we must all agree, all work together, we're no better than a machine. If an individual can't work in solidarity with his fellows, it's his duty to work alone. His duty and his right. We have been denying people that right. We've been saying, more and more often, you must work with others, you must accept the rule of the majority. But any rule is tyranny. The duty of the individual is to accept *no* rule, to be the initiator of his own acts, to be responsible. Only if he does so will the society live, and change, and adapt, and survive. We are not

---

[19] Zygmunt Bauman, *Postmodern Etik*, trans. Alev Türker (İstanbul: Ayrıntı Yayınları, 2020), 136.
[20] Seran Demiral, *Ursula K. Le Guin* (İstanbul: Gerekli Kitaplar Yayınevi, 2019), 45.
[21] Le Guin, *Mülksüzler*, 295.

objects of a State founded upon law, but members of a society founded upon revolution.²²

With this idea in mind, Shevek suggests that all the ideas that emerge in Anarres are quite lacking in the point of practice. Odonian society, which covers the true meaning of freedom and revolution, has failed in this regard of the will to power. Seemingly, it reflects the ideal society in which individuals have the right to choose in each field according to their own capabilities and wishes. But how can one explain, then, that Shevek's important works on time matter are not allowed to be published and that the reward from Urras is concealed when those works are published under certain conditions? At this point, a different question arises that Le Guin evokes. Are freedom, free will, the right to choose, and the right to progress for everyone or private? While Le Guin does not explicitly answer this question, she arouses this suspicion in the reader by allowing these thoughts to pass through Shevek's mind. Moreover, when Shevek is certain that these thoughts are painfully correct and that there is no free will in Anarres anymore, he experiences enlightenment. He questions why this was allowed to happen in Anarres. What is the driving force behind this renouncement? As stated before, scarce resources, social pressure, fear of isolation, and, most importantly, the anxiety of being different from others can be listed among the reasons for the abandonment of free will. But the different awakening point is that Shevek expresses this verbally to someone else for the first time. After the thoughts in his mind are expressed, the response from the universe and the justification of this awareness are not delayed. Sadik, daughter of Shevek and Takver, like all other children in Anarres, stays at night in the dormitory, which is one of the rules of Anarres society. For this Odonian society, where social solidarity and sense of unity are at the forefront, it is of course quite common to have many common areas of use. But the question is, do these common areas of use increase unity and solidarity, or are they places that offer opportunities to isolated individuals who want to experience their right of choice? That night, Shevek learns the answer to this question from Sadik's mouth:

> Sadik answered him with desperate courage. 'Because they don't like—they don't like the Syndicate, and Bedap, and—and you. They call—The big sister in the dorm room, she said you—we were all tr—She said we were traitors,' and saying the word the child jerked as if she had been shot, and Shevek caught her and

---

²² Ibid., 305-306.

held her. She held to him with all her strength, weeping in great gasping sobs.[23]

The society itself needs to be protected against any kind of initiation that will damage its basic principles. Seen another way, while freedom is one of the fundamental principles of that society, is it reasonable to label individuals as traitors because they choose freedom? As Jaeckle asks, "How can one ensure that the acts she initiates will always be in the best interests of the group?"[24] Throughout the novel, Le Guin concentrates upon this dilemma. Between the lines, with critical reading, this idea comes to mind: Freedom is never too safe. If you choose it, you can lose your dignity, mutuality, and sociality to some extent. In the simplest terms, if you decide out of fear of not losing all this, may there be talk of free will? In this context, it is possible to question Shevek's decision about going to Urras to study and share information for the benefit of all planets. One can ask why he chooses to go there; is it really necessary if it is against Odonian rules. In a climactic paragraph, Shevek's resolution to identify his desire becomes clear:

> Here I'm walled in. I'm cramped, it's hard to work, to test the work, always without equipment, without colleagues and students. And then when I do the work, they don't want it. Or, if they do, like Sabul, they want me to abandon initiative in return for receiving approval. They'll use the work I do, after I'm dead, that always happens. But why must I give my lifework as a present to Sabul, all the Sabuls, the petty, scheming, greedy egos of one single planet?[25]

While this confession, and the question that comes after it, clearly explains Shevek's reason for going to Urras, it also criticizes the practices going on in Anarres. "Le Guin dreams of utopian worlds and moons, free of the inequalities of capitalism and the injustices of state power. More importantly, however, Le Guin develops new forms of anarchist thinking,"[26] writes Call to indicate the novel's engagement of the critical mind via Shevek's questioning. This questioning can be accepted as another kind of anarchist thinking.

---

[23] Ibid., 315.
[24] Jaeckle, *"Embodied Anarchy,"* 79.
[25] Le Guin, *Mülksüzler*, 322.
[26] Lewis Call, "Postmodern Anarchism in the Novels of Ursula K. Le Guin," *SubStance* 36, no. 2 (2007): 88, https://doi.org/10.1353/sub.2007.0028.

Conclusion

There are several pathways that genuinely imply the disparity between deterministic grounds and free will in the novel. In seeking to persuade us of this, Le Guin forces Shevek to make a choice, which is against the social norms of Odonian society. With this choice—more clearly, with his own wish—Shevek exemplifies the situation that is out of control and incompatible. Le Guin describes the physical conditions and philosophical aspects of this planet in detail to persuade readers that collective consciousness and mutual aid are vitally important to survive there, so it can be expected that author supports this kind of determinism. On the contrary, closer examination reveals that she simply wants to make it clear that even in an ungoverned state that is founded on freedoms and personal rights, egoist views or hierarchical organisms that have the power to hinder free will may exist. Some grounds can be found for rejecting this controversial issue, of course; however, Shevek's confessions amalgamated with different examples that appear on Anarres provide a way to restructure the problem. Underlying physical factors as determining forces upon free will restrict libertarian socialism in Anarres. Somewhat surprisingly, there is also a counterfactual situation that restricts freedoms: the nature of human beings. Le Guin states a quarrel with her essential hypothesis of why it is not possible to decide via free will in a society that is considered supportive of freedoms. And she suggests just one single path that can be evaluated as the true answer for this quarrel. For the sake of the argument, that human factor surpasses physical conditions in terms of choices; Le Guin criticizes causal and non-causal factors in terms of their efficiency.

References

Bauman, Zygmunt. *Postmodern Etik.* Translated by Alev Türker. İstanbul: Ayrıntı Yayınları, 2020.

Bergson, Henri. *Time and Free Will.* New York: Dover Publications, 2016.

Bracke, Astrid. "Flooded Futures: The Representation of the Anthropocene in Twenty-First- Century British Flood Fictions," *Critique: Studies in Contemporary Fiction* 60, no. 3 (February 2019): 278-288. https://doi.org/10.1080/00111619.2019.1570911.

Call, Lewis. "Postmodern Anarchism in the Novels of Ursula K. Le Guin," *SubStance-The John Hopkins University Press* 36, no. 2 (November 2007): 87-105. https://doi.org/10.1353/sub.2007.0028.

Demiral, Seran. *Ursula K. Le Guin.* İstanbul: Gerekli Kitaplar Yayınevi, 2019.

Fischer, John M. *The Metaphysics of Free Will.* Oxford: Blackwell Publishers, 1994.

Garrard, Greg. *Ekoeleştiri. Translated by* Ertuğrul Genç. İstanbul: Kolektif Kitap, 2020.

Jaeckle, Daniel, P. "Embodied Anarchy in Ursula K. Le Guin's The Dispossessed." *Utopian Studies* 20, no. 1 (2009): 75-95. http://www.jstor.org/stable/20719930.

Le Guin, Ursula K. *Mülksüzler.* İstanbul: Metis Yayınları, 2020.

Smart, Jack. J. C. "Free-Will, Praise and Blame." *Mind* 70, no. 279 (1961): 291-306. http://www.jstor.org/stable/2251619.

Trexler, Adam. *Anthropocene Fictions: The Novel in a Time of Climate Change.* Charlottesville: University of Virginia Press, 2015.

Van Inwagen, Peter. "When is the Will Free?" *Philosophical Perspectives* 3, (August 1989): 399-422. http://www.jstor.org/stable/42920381

CHAPTER FIVE

THE POST-APOCALYPTIC WORLD ORDER IN *THE TIN CAN PEOPLE* BY EDWARD BOND[1]

ELVAN KARAMAN

The trilogy of *The War Plays* by Edward Bond offers an "envisioning of nuclear disaster."[2] The plays in the trilogy are "harrowing and controversial" as they "chronicle the reasons behind the outbreak of nuclear war, the impact on those who survive and the ultimate consequences of survival."[3] As Rabey suggests, "Bond's appetite for morally paradoxical dynamics – whereby the surprising, conventionally 'wrong' action throws into relief the ultimately worse evil of the generally accepted social 'order' around it – is continued in the trilogy *The War Plays*."[4] This study will analyse the second play, *The Tin Can People*, where "Bond suggests that we may have to suffer a nuclear holocaust to expose the injustices of our day."[5] *The Tin Can People* is a post-apocalyptic play that depicts "an image of survival after nuclear destruction," including the condition of the world and the survival of only a few characters after a nuclear explosion.[6]

---

[1] This article has been adapted from Elvan Karaman's unpublished PhD thesis, entitled "Theatre as a Product and a Mirror of Socio-Economic Structure in English Society." Specifically, it is derived from pages 153-77.
[2] Jenny S. Spencer, *Dramatic Strategies in the Plays of Edward Bond* (Great Britain: Cambridge UP, 1992), 222.
[3] Barry B. Witham, "English Playwrights and the Bomb," *International Journal of Politics, Culture, and Society* 2, no. 2 (1988): 297, http://www.jstor.org/stable/20006902.
[4] David Ian Rabey, *English Drama since 1940* (Great Britain: Pearson Education, 2003), 84.
[5] Robert D. Hostetter, "'Drama of the Nuclear Age': Resources and Responsibilities in Theatre Education," *Performing Arts Journal* 11, no. 2 (1988): 94, http://www.jstor.org/stable/3245646.
[6] Janelle Reinelt, *After Brecht: British Epic Theater* (USA: The University of Michigan Press, 1996), 60.

The criticism of modern people who have caused a catastrophe with a capitalist mindset is apparent throughout the play. The damage to nature, animals and people occurs as a result of human beings' capitalist greed for money and power. Similarly, the scientific knowledge that people have studied for decades is criticised as people are deprived of self-creation while they develop science. As a result, they revert to pre-historic circumstances while trying to improve the world and their conditions. This problem has two significant consequences. People become alienated from nature around them and their own nature, and they also lose their ethical values and become conditioned to be egocentric and evil. They behave like people who were 'in the state of nature' in ancient times. In Reinelt's words, "The Tin Can People think they can build a paradise on the ruins of the old society, but they are inscribed in the codes and practices of the dead social order and reenact those patterns."[7]

The Criticism of Modern People

In *The Tin Can People*, Bond has created an influential and often shocking discourse with which he criticises modern people for having a number of negative features such as being selfish, ambitious and cruel. In addition to the negative traits of human beings, the capitalist system also supports individualism and the will to obtain wealth as well as profit in modern people instead of a desire to improve one's character. The consequence makes people greedier and harsher at each step as the system and its ideology do not allow them to realise the vicious circle and step out of it. Although there are certainly some people who reject the order of the capitalist system, most people are pushed towards increased profit at the expense of others. In this respect, Bond, an "anti-authoritarian [and] utterly committed to the relief of human suffering and to the alleviation of man's cruelty to man," critiques this situation in his play and underlines the high probability of being drawn into an eventual catastrophe for the whole world.[8] For example, as modern people have been developing science for more than a century, they have seen that they are able to do anything with their reason and ambition. Developing science also means producing, earning and consuming more and being more powerful as a consequence of the capitalist drives of people without thinking about the future of humanity on earth. Hence, Bond criticises this greedy and

---

[7] Ibid., 60-1.
[8] Michael Patterson, *Strategies of Political Theatre: Post-war British Playwrights* (Great Britain: Cambridge UP, 2003), 410.

reckless attitude and emphasises its probable outcomes. As Daneshzadeh argues, "Bond has repeatedly questioned the rationality of a society 'in which the audience can sit in the theatre while a few miles along the road men are sitting before gadgets that fire nuclear weapons' and he said that he wanted 'to show the psychosis that exists after a nuclear war'."[9] The Second Chorus states:

> The world was made into a crucible for an experiment
> The effects couldnt have been foreseen
> We called them the voice of the bomb
> It spoke everywhere: we dont know how its orders were put into effect.[10]

Hence, the explosion of the bombs during the experimentation of some scientists almost leads to the end of the world. As Patterson observes, "Wherever the future lies, Bond fears there is not much time left to steer towards it. Given the achievements of modern technology, whereby humankind faces imminent self-annihilation from nuclear weapons or through destruction of the environment, we are, in Bond's view, teetering on the edge of an abyss."[11] The First Chorus accordingly describes how the world looks after the unnatural disaster:

> Years later a dust as white as old people's hair settled on everything
> The world looked like a drawing in lead on white paper
> Hours after the explosions I walked over a bridge
> The thirst caused by the fires was so severe that even the drowning called for water
> People fled in all directions from one hell into another.[12]

There is chaos even years after the explosions and the white dust covering everything shows that the effects of the disaster are still fresh. Bond emphasises this reality with his unexpected and grotesque dramatisation.[13] As Daneshzadeh observes, "*The War Plays*' trilogy…presents the scenario of a waste land with apocalyptical shades. The post nuclear environment of the plays reflects the Atmosphere of the historical period when [they were] written. The beginning of the eighties saw the debate

---

[9] Amir Daneshzadeh, "Analysis of Edward Bond's *War Plays*," *International Letters of Social and Humanistic Sciences* 61, (2015): 2, https://www.scipress.com/ILSHS.61.1.pdf.
[10] Edward Bond, *The War Plays: A Trilogy* (Great Britain: Methuen, 1985), II.40.
[11] Patterson, *Strategies*, 410.
[12] Bond, *The War Plays*, I.33
[13] Witham, "English Playwrights," 298.

about nuclear weapons."[14] As a result, Bond argues that the entire world has been transformed into hell by the ambition of modern people.

Bond makes readers face the future they have long been afraid of experiencing. As Spencer points out, "Set in a vaguely futuristic time after 'rockets destroyed the world,' one of the plays' first challenges is to render...the 'unimaginable' or 'unthinkable' horror of nuclear destruction, its causes and probable effects."[15] Hence, human beings are responsible for bringing about the end of the world. As the Second Chorus states:

> Skeletons sat before stone buttons and stone computers
> Stone politicians and stone officers bent over stone maps of cities they had made dust
> Our ancestor who sheltered in caves painted the walls with scenes of life
> These had covered them with charts of death
> They were like pharaohs who killed their servants.[16]

Not only the electronic devices but also the people around them have been transformed into stone due to the greedy experimentations of the governors. Bond also makes a remarkable connection between the present time and the past. In Spencer's words:

> [T]he Second Chorus...introduces an important connection between history and humanity, scientific knowledge and self-creation. On the one hand, our own age, looked back upon from the future, may seem inhuman; the metaphor is a warning. But more importantly, the frozen relic metaphor reminds the viewer that prehistoric time is not, strictly speaking, human time; we cannot go forward by going back.[17]

In the competitive atmosphere of the Cold War, during which the trilogy was created, everyone knew that many countries were working to develop different kinds of bombs, such as nuclear and hydrogen bombs. In other words, the struggle to be stronger and wealthier was being waged not only individually but also internationally. People were also aware of the possible results of this competition for power. In that political climate, Bond could not be optimistic since he believed that the fuse of nuclear war

---

[14] Daneshzadeh, "Analysis of Edward Bond's *War Plays*," 1.
[15] Spencer, *Dramatic* Strategies, 222.
[16] Bond, *The War Plays*, II.41.
[17] Spencer, *Dramatic Strategies*, 235.

had already been lit and that calamity was inevitable.[18] Therefore, he wanted to alert people and he created his own discourse in which a nuclear disaster destroys the world and nearly everyone in it.

Damage to Nature and Human Beings

People's ambition to produce more is constantly stimulated by the capitalist system and this ambition results in the consumption of more of the planet's natural resources, because of which people now face a climate crisis. This attitude has already started to cause a number of environmental problems and this man-made disaster is damaging nature. Bond's characters witness how nature has died as a result of nuclear bombs. The Third Woman states: "no doubt it had been made by a bomb – and I walked over it – the heat had turned the ground to glass / ... So I said die die die / ... in fact it had become the greeting in that city / I told everything I saw to die."[19] Bond highlights that the influences of a probable catastrophe can ruin not only human beings but also nature, to which people belong. As a result of these experiences, "die" was the only word that the Third Woman uttered for a long time. The Second Man also maintains: "We know what hell is – the fires – the shock wave: like being shoved by a team of giants – the long winter: clouds suddenly froze solid and ice blocks fell and crushed many who'd survived."[20] Thus, the bomb explosions destroyed nature so severely that they caused extraordinary weather events, and human beings are helpless and weak when nature reacts to human actions. Although people rarely care about it while they are greedily experimenting or developing science and technology, they cannot escape from the terrible impacts of damaging nature as human society is "controlled by violence and repression, threatened by science and technology."[21] In other words, human beings are a part of nature and they damage themselves while they are damaging nature for their own profit, as is clearly seen in the effects of climate change today.

As a part of nature, animals have also been exterminated by the catastrophe. The Second Man, for example, asks: "Why did I think of sheep on a hill? / There's no green grass now."[22] Hence, Bond is among the writers who have "provided even darker visions of the future

---

[18] Witham, "English Playwrights," 297.
[19] Bond, *The War Plays*, I.ii.35-6.
[20] Ibid., I.ii.35.
[21] Daniel R. Jones, "Edward Bond's 'Rational Theatre'," *Theatre Journal* 32, no. 4 (1980): 506, http://www.jstor.org/stable/3207412.
[22] Bond, *The War Plays*, I.ii.37.

consequences of Cold War tensions, especially...nuclear."[23] The First Man also observes that dead bodies of animals are as plentiful as corpses of people as they were all affected together: "No hares or birds or rats / Bones of animals and people lie together on the road like broken picture frames / The flies flew off: great swarms: I thought the sky was full of funerals."[24] Thus, dead bodies of people and animals lie together even though people had already cut their connections with animals in the harsh competition of the consumerist system. The perspective of the world is not promising and to continue living among millions of dead people and animals seems very hard for the survivors. In addition, the soil expels fossilised animals and people, which makes the scene even more horrible together with the bodies of the newly deceased. Bond depicts here the possible appalling influences of experiments for new and more effective nuclear bombs as *The War Plays* "are set in the aftermath of a nuclear holocaust, tracing the subsequent actions of survivor communities."[25] The plays "offer moments of startling and memorable poetic imagery."[26] Since Bond is conscious of the fact that "only extremity may produce (or necessitate) real change in human behaviour," he pulls the audience and the reader into an extreme dramatic experience so as to encourage "an awareness of the mind-numbing ubiquity of the problem" and the urgent necessity of taking action to solve it before it is too late.[27] He struggles to make a call to all people to do something in order to prevent this horrible fate: "Acceptance is not enough ... you can sit quietly at home and have an H-bomb dropped on you. Shakespeare had time ... But ... for us, time is running out."[28]

The nuclear experiment annihilates almost the entire population of the world, as well. The destruction of the bombs is so extensive that there are only seven people remaining on the earth, and that number is also decreasing throughout the play. Bond employs these survivors as eyewitnesses to the disaster. The Third Woman describes it as follows:

> How can you talk about the destruction of the world and be normal? ...

---

[23] Keith M. Booker, *The Dystopian Impulse in Modern Literature: Fiction as Social Criticism* (USA: Greenwood Press, 1994), 172.
[24] Bond, *The War Plays*, I.33.
[25] Rabey, *English Drama*, 84.
[26] Ibid., 84.
[27] Spencer, *Dramatic Strategies*, 224.
[28] Patterson, *Strategies*, 414.

> I tried to help the wounded – there was no medical knowledge or drugs
> The sick came together in a few places – crawled and limped along the streets – followed each other's cries
> When the blind touched the walls they fell on them
> I shouldnt have started to speak
> In the mornings there were rooms full of the dead.[29]

Their experiences are horrible; they have had to see the agony of people suffering awfully around them. As millions of people were impacted by the disaster, there were not enough of them to help each other. This increased the pain that people had to bear and the world was no different from hell. All these experiences were traumatic for the survivors. The Third Woman remarks in agony: "I cant stop speaking / Of course babies suckled their dead mothers and mothers tried to give milk to their dead babies."[30] Bond emphasises how human beings can create an unbelievable disaster, the outcomes of which no one can imagine. In the 1980s, people were also dismayed by the probable future results of a nuclear bomb. In Carter and McRae's words, "There was a veritable explosion of expression around the question of the atomic bomb, around the possibility that all life could end at a moment's notice."[31] Similarly, the results of the man-made disaster in Bond's play are so horrific that they still impact the characters after seventeen years. What is more, each one's experience is worse than the others and they cannot escape from the psychological influences of the event:

> FIRST MAN. In the desert I found a skull eating a stone
> Been clenched in its teeth to stop the pain
> I shouted – a cliff – wind blew the echo back through the skull's mouth
> It spoke
> A bit higher – a woman's voice
> Slept beside her in the sand
> One of the rare dews: in the morning her cheeks were wet.[32]

How much sorrow people suffered due to this nuclear holocaust is apparent. While people imagined that they had developed beyond belief,

---

[29] Bond, *The War Plays*, I.ii.35.
[30] Ibid., I.ii.35.
[31] Ronald Carter and John McRae. *The Routledge History of Literature in English: Britain and Ireland*, 2nd ed. (New York: Routledge. 2001), 366.
[32] Bond, *The War Plays*, I.ii.38.

the point they reached is like the beginning of history again with few people on the earth. Nevertheless, it is worse this time, because human beings have neither nature nor animals around them. Above all, the characters do not have names, but only numbers, because it is a new beginning for the world and everything has changed. The world order that existed before the explosions is gone and the characters do not have identities, including families or nationalities. As Bond notes, "They have lost their names because they have lost themselves. Names are a sign of our humanity. In a nuclear age we still have to create our humanity."[33] On that account, creating nameless characters is another way of depicting "a horrific world in which humanity has disappeared after the nuclear disaster."[34]

Scientific Knowledge and the Lack of Self-Creation

Human beings' struggle to improve science and technology dates back to the Industrial Revolution. People have always sought to transcend their time with scientific and technological developments. However, scientific knowledge brings about a number of drawbacks for people, such as cutting the bond between human beings and nature. In this respect, people only achieve physical developments, such as creating new factories, smart devices, robots and artificial intelligence. They cannot improve the essential traits for humanity and they are deprived of self-creation in their technological world. The man-made disaster that people caused while trying to develop a new kind of bomb is also an outcome of this deprivation and human beings have been returned to pre-historic times. As Reinelt mentions, "The third section leaves the survivors at the beginning of their attempt to start over and learn how to create a changed future. ... They will try to learn from their mistakes and from the things they know about the past. If they survive, future generations will look back on them, 'and bombs'll be as old to them as stone axes are to us'."[35] Nevertheless, the survivors cannot achieve a fresh beginning; they continue making similar errors. Bond thus argues that pre-historic conditions are inhuman and not suitable for human beings. This can be taken as a warning for people to be careful about their relationship with science and the probable results of developing it without limits.

---

[33] Daneshzadeh, "Analysis of Edward Bond's *War Plays*," 4.
[34] Ibid., 4.
[35] Reinelt, *After Brecht*, 62.

The bond between nature and human beings is one of the factors that contribute to the self-creation of people. However, the development of scientific knowledge and the machinery produced by this knowledge itself have progressively weakened that bond. Over time not only urban people but even farmers have fully lost their bonds with nature and become imprisoned by modern living conditions, surrounded by technological devices. In consequence, people have also been alienated from their own nature. As Bond himself suggests, "We have technological problems—the machines we make are too powerful for us. Instead of, as they did in the past, enabling us to improve our relationship with nature, they now damage our relationship to nature. And so whereas tools were the makers of humanness now tools are becoming anti-human."[36] In the play, the survivors are faced with pre-historic conditions after the explosions. Nature is dead around them, so they do not have any opportunity to grow vegetables or fruit. There are no animals alive, either. Hence, the survivors do not have the possibility of establishing relations between producer and production without nature and animals around them. The survivors are therefore more primitive than the people who lived in the pre-historic past. The Tin Can People have no devices to support them in their primitivity. They are still alienated and cannot improve themselves under these conditions. For example, the only food they can eat is food in tins, as everything else has been destroyed. The food in the tins, which sounds unnatural and synthetic, reminds the reader of the fast food that has been popularly eaten since the late twentieth century, and the characters are obliged to eat this food. As people are what they eat, Bond names the survivors "tin can people," because they also lack some significant human features such as emotions. At the same time, they still have most of the negative features of the previous culture. Reinelt writes as follows: "The tin cans, which are stockpiled in ware-houses, insure that this new society will not suffer the scarcity of the old, but they are also the visible sign of the previous culture, with its emphasis on possession and greed."[37] In other words, Bond stresses their continuing alienation and consumerism. As Rabey explains, "*The Tin Can People* shows a community given apparent 'labourless luxury outside financial time' by discovery of an abandoned warehouse full of tinned food. However, their consumerist bliss contains the seeds of violence in their fearful reflex to preserve it at the expense of outsiders."[38] Correspondingly, their eating so much catches

---

[36] Peter Billingham, *Edward Bond: A Critical Study* (London: Palgrave Macmillan, 2014), 3.
[37] Reinelt, *After Brecht*, 61.
[38] Rabey, *English Drama*, 84.

the attention of the reader and the audience, which is not new for people in the twenty-first century. "[T]he extreme of consumerism" with the "[s]tarving people gorging in the ruins" is emphasised here, as well.[39] Therefore, the survivors go about their lives as they were taught in capitalist society since they cannot produce anything or develop humane relations of production.

Losing Ethical Values

In *The Tin Can People*, Bond observes that human beings have only developed the world around them, not themselves. Their egos are still extremely rudimentary and they still have their primitive fears and prejudices. As Reinelt puts it, "The Tin Can People think they have built paradise 'in the ruins of hell,' but they are still in the grip of the fear of scarcity and their notions of prosperity still come from the previous destructive era."[40] Furthermore, they become much more violent without the control of institutions such as education, religion and law. Therefore, the Latin proverb *Homo homini lupus est*, which means "A man is a wolf to another man," becomes true for the survivors. As described by Thomas Hobbes, they are living in the state of nature and always trying to destroy each other in order to protect themselves and their own benefits. The survivors consider the world after the catastrophe to be a paradise, because there is no need to work. They have much more food, housing and land than they need. The earth that they have inherited is too big for seven people and their lives are too simple to have any reason to fight. However, they do not fail to turn their paradise into hell with their prejudices and the teachings of the capitalist system, which they begin applying again at the first possible chance. In Witham's words, "These survivors, after many years, carve out a rudimentary society which they again destroy in riots and panic following the arrival of a stranger whom they believe might be a plague carrier."[41] As a result, human beings cannot annihilate their prejudices even if they nearly destroy the whole world.

The instinct of protecting oneself is seen in most of the characters. For instance, when the First Man, having survived alone so far, encounters the First Woman, he feels unsettled as he is afraid of being attacked or murdered when she offers him some food in a tin. The other survivors also have a similar distrust of the First Man, wanting to protect themselves as a

---

[39] Daneshzadeh, "Analysis of Edward Bond's *War Plays*," 4.
[40] Reinelt, *After Brecht*, 55.
[41] Witham, "English Playwrights," 297.

number of diseases have been spreading since the nuclear explosion.⁴² In this respect, the survivors live in a state of nature, do not rely on each other, and believe that they always need to be careful. In fact, the First Man is proved to be right in a short time after he joins the other survivors. While he merely wants to protect himself, driven by an innocent instinct, the other survivors have a strong desire to murder him so as to eliminate the possibility of death from the group. As Daneshzadeh maintains, "When a stranger appears, he is welcomed into the group, but suspicions mount when one of the other survivors dies. Convinced that the newcomer is contaminated, the group resolves to destroy this new threat to their existence."⁴³ Here, Bond highlights the struggle of the group to cast out and destroy the newcomer with their prejudices and primitive fears as they "reenact the behaviours of the dead culture toward 'difference'."⁴⁴

The capitalist system has transformed human beings, especially the good in them. As Innes argues, "For Brecht, personality is not innate, but determined by social function. The exploitive class system of capitalism imposes the strain of being evil on men, making the natural instinct to goodness a fearful temptation to be avoided because self-destructive. Bond has restated Brecht's viewpoint almost exactly."⁴⁵ The survivors, for instance, are extremely egotistical and are happy to see the First Man as they can use him for their own benefit. The Third Woman remarks: "I bet he runs like the wind! / A person! Alive! / We'll get him to run for us!"⁴⁶ Bond criticises the manner of modern capitalist people here as the survivors continue behaving according to the teachings of that system. Although the First Man could have been a chance for them to make a good, new start, they choose to be evil.

Thus, the survivors are blinded by their fears and bias against the First Man when the Third Woman suddenly dies a short time after his arrival. Although they lack even the slightest evidence, they immediately decide to kill him to make sure that they will go on living. There is a narrow-minded resistance and prejudice against a new person in the group. As they are not overseen by the authority of any state or institutions, they become more reckless in their behaviours towards the First Man and later

---

⁴² Daneshzadeh, "Analysis of Edward Bond's *War Plays*," 4.
⁴³ Ibid., 5.
⁴⁴ Reinelt, *After Brecht*, 61.
⁴⁵ Christopher Innes, "The Political Spectrum of Edward Bond: From Rationalism to Rhapsody," in *Contemporary British Drama, 1970-90: Essay from Modern Drama*, ed. Hersh Zeifman and Cynthia Zimmerman (London: Macmillan, 1993), 89.
⁴⁶ Bond, *The War Plays*, I.ii.37.

each other. Although one or two of the survivors sometimes doubt the necessity of killing him, they frighteningly convince each other. Bond critiques the way in which people exclude a helpless individual even under these conditions; this prejudiced attitude is similar to a genetic disease among human beings. Fights start again as all of the other survivors accuse the First Man of having a contagious disease. He faces the accusations in shock and horror:

> SECOND WOMAN. (*points to the* FIRST MAN): Its him
> FIRST MAN. (*unsure*) Me? – what? – you said no trial –
> SECOND MAN. You met people on your way here –
> FIRST MAN. People?
> SECOND MAN. Dont argue – you must have – at the beginning Did they die? Did they fall down like this?
> SECOND WOMAN. He's on the run! He's been turned out! He's dangerous![47]

They treat him with discourteousness and quickly start planning to kill him. "In the *Tin Can People*," Witham argues, "a group of survivors explode into an orgy of ransacking and killing because of the fear generated by an 'outsider' in their midst. Although he is young and possibly capable of siring children in the barren population, the stranger is not perceived in 'human' terms."[48] Bond here discusses not only the violence in the nature of human beings but also their limitlessness when they believe that their own lives or profits are at stake. To Castillo, "Bond's theater is political, and his focus of investigation is nothing less than the survival of all the human, humane qualities of the political animal, the dweller in a contemporary polis."[49] Therefore, the survivors behave like wild animals without any authority over them. They have a "moral and political bankruptcy that once 'attained', cannot be easily overturned."[50]

Their plans to kill the First Man are also very insidious and violent. The process of their attempt to kill him is more shocking at each step. As Daneshzadeh suggests, "For Bond, the violence against the socially marginalized portrays the unjust system of our world."[51] First, they claim that they will put him in quarantine, but they ultimately intend to kill him. The First Man is the innocent lamb here and begs to stay with them: "Let

---

[47] Ibid., I.ii.40.
[48] Witham, "English Playwrights," 298.
[49] Debra A. Castillo, "Dehumanized or Inhuman: Doubles in Edward Bond," *South Central Review* 3, no. 2, (1986): 78, http://www.jstor.org/stable/3189368.
[50] Reinelt, *After Brecht*, 69.
[51] Daneshzadeh, "Analysis of Edward Bond's *War Plays*," 4.

me stay with you and my journey will've been a green track over the desert / Fetch your water – open your tins / All I want is – look at your faces – speak."[52] However, the others are so blinded by their prejudices that they do not hear him. Hence, they start to talk about the ways in which they can meet their aim:

> SECOND MAN. We'll have to kill him
> SECOND WOMAN. How?
> We cant get close enough to do it with a knife or hammer – he'd run for it
> There arent enough of us to stone him
> When they stoned people it took a whole town
> FIRST WOMAN. We mustnt panic – try to think!
> SECOND WOMAN. I feel so helpless
> It was easy for the bombs to kill millions!
> How d'you kill just one?[53]

Bond emphasises the severity of the cruelty in people's hearts with these sentences; in particular, the Second Woman even wishes for bombs to kill the First Man after having survived such a disaster herself resulting from the explosion of bombs. As Daneshzadeh argues, "Whilst the cold war and the fin-de-siècle have now passed, the play still has an eerie resonance. We are still in a state of paranoia - post 9/11, post 7/7, in a world where military intervention is a confirmed reality."[54] In that regard, the Second Woman's sentences reflect the history of humanity, which has not had the slightest improvement in its characteristics. People have also lost their ethical values and have the potential to treat each other with any kind of violence. The Second Woman makes the others feel comfortable because there is no authority to punish them: "If anyone came looking for him they'd never even find his body in this charnel house / We live in dead people's clothes – eat their food – we took the storekeys from dead soldiers' pockets / One more wont make the skeletons cry."[55] As she supposes that there will be no penalty, she feels free to commit a crime and murder an innocent man. In Witham's words, "[M]ore importantly, the holocaust will not end the slaughter or the suffering. Even after the missiles fly and the survivors put their crumbled world back together, they are still susceptible to the same deadly arms race if they do not fashion

---

[52] Bond, *The War Plays*, I.ii.38.
[53] Ibid., II.i.41
[54] Daneshzadeh, "Analysis of Edward Bond's *War Plays*," 4.
[55] Bond, *The War Plays*, II.i.42.

societies which are truly just."⁵⁶ Consequently, they have lost their ethical values and do not learn their lesson from the nuclear holocaust that they have experienced. They continue their violent attempts to destroy each other.

The Second Man, volunteering to kill the First Man, puts an end to the debate about annihilating their victim and the level of violence rises incessantly as he suggests various materials to achieve their aim. The cruelty and evil in people's hearts come to the surface and become more horrible with each passing moment. As Castillo puts it, "Where the traditional fables of Aesop use animals to demonstrate human foibles, Bond's modern fables utilize a similar, if more subtle, version of this technique by surrounding characters with images of dehumanization or metaphors connecting them to animals."⁵⁷ On that account, the violence of the survivors proves that they are still wolves to one another and that human beings cannot change or develop over time. As Bond emphasises, "I'm not interested in violence for the sake of violence. Violence is never a solution in my plays, just as ultimately violence is never a solution in human affairs. Violence is the problem that has to be dealt with."⁵⁸ Hence, violence in the plans of the survivors shows the severity of the problematic nature of human beings that does not disappear. The Second Man makes detailed plans before the murder: "When I've killed him I'll break the spear and live on my own for six months / Not out of guilt: he has to be killed for the community's sake / Six months will be a sign of respect we owe all the dead / And it'll show anyone who heard we'd killed him that we did it reluctantly / After six months you'll come and welcome me back."⁵⁹ The murder, on which they all agree, is justified as they try to find excuses to kill him. Thus, their crime turns into a ceremony of murder. The Second Man's plan is extremely realistic as he is similar to modern people. According to Spencer, "[T]he 'Second Man,' who so eloquently articulates the hopes and beliefs of the group, not only reinvents (with the logic of our own political leaders) a weapon for their protection, but offers to 'sacrifice' himself in their defense. [This is the] recognisable logic and emotional impulses of the members of the group, so clearly projected from our own society..."⁶⁰ As a result, people will never learn from their faults

---

⁵⁶ Witham, "English Playwrights," 298.
⁵⁷ Castillo, "Dehumanized or Inhuman," 78.
⁵⁸ Karl-Heinz Stoll, "Interviews with Edward Bond and Arnold Wesker," *Twentieth Century Literature* 22, no. 4. (1976): 415, http://www.jstor.org/stable/440383.
⁵⁹ Bond, *The War Plays*, II.i.43.
⁶⁰ Spencer, *Dramatic Strategies*, 232-33.

and fights will always continue as long as human beings exist on earth, for they always "[re-enact] the destruction they originally escaped."[61] As Bond himself suggests, "The simple fact is that if you behave violently, you create an atmosphere of violence, which generates more violence … So a violent revolution always destroys itself."[62]

Bond, therefore, underlines how cruel human beings can be and shocks the reader with the Second Man's sudden death while he is planning the murder. The playwright punishes not only the Second Man but also the other survivors. He does not allow the Second Man to murder the innocent First Man; Bond prevents the murder of Abel by Cain this time. Above all, he kills Cain to teach a lesson to the reader. He emphasises that people will not manage to survive in such a world this time if they do not change their mentality and wild characteristics. As Howard observes, "The Tin Can People are more than Bond's image of suicidal materialism; the play is given nauseating weight by their language, for it dwells remorselessly on the past; they are fixated with the dead and the physical horror of the catastrophe."[63] Accordingly, Bond creates these characters to show human beings their wild nature and the nightmarish future that will result from it, reminding them how to return back to humanness again. He argues in his interview with Billingham: "The only way that you can create humanness is by dramatizing the self. We should be dramatizing the conflicts within the self and what art and drama should be doing is increasing human self-consciousness. … Once you engage in that process you have to start asking, why am I committed to humanness?"[64] Hence, the degenerated behaviour of the characters is employed by him to shock people and remind them how they should really be. He challenges the reader and the audience to realise their corruption and "to accept responsibility through action."[65] The only possible change will occur with the right-minded action of human beings again.

Conclusion

Based on the above analysis, the capitalist mentality of the survivors and the limitlessness of their evil are criticised in *The Tin Can People*. As Bond points out: "One can talk about the culture of socialist man but I'm not talking about culture, I'm talking about humanness.

---

[61] Ibid., 232-33.
[62] Patterson, *Strategies*, 414.
[63] Daneshzadeh, "Analysis of Edward Bond's *War Plays*," 4.
[64] Billingham, *Edward Bond*, 3.
[65] Jones, "Edward Bond's 'Rational Theatre'," 517.

Culture will sustain itself but humanness must be re-created."[66] However, the survivors are far removed from re-creating humanness. Bond argues that human beings in the capitalist system always carry hell and "the legacy of the past" in their hearts even though they suppose that the world is heaven for them now.[67] The aforementioned Latin proverb seems to be true throughout the play and the capitalist mindset encourages people to be more egotistical and violent for their own benefits. In Jenkins' words, "Bond ... envision[ed] an apocalyptic future. *The War Plays* ... moved beyond narrow politics. War turns men to beasts and the earth to ashes; worse still, the few survivors begin to repeat old patterns of fear and aggression."[68] Bond means to show us that if people go on living like beasts even when they are returned to pre-historic times, their ongoing faults will deprive them of any future. Therefore, his trilogy is a strong warning for humanity to abandon its primitive and prejudiced mindset. Otherwise, the world will be transformed into hell for them. The Fourth Woman summarises the journey of human beings in the world: "We dont learn from other people's mistakes – not even from most of our own. But knowledge is collected and tools handed on. We cant go back to the beginning, but we can change the future."[69] This is the gist of the play and all disasters happening for centuries. As long as people let themselves be shaped by and obsessed with violence and ambition, they will not have a future.[70] Thus, Bond "fuses the future with the present" in his trilogy.[71] The forthcoming apocalyptic future of human beings is not distant for twenty-first-century people who have been experiencing the impacts of the climate crisis alongside the pandemic, but a considerable number of them are still far removed from any consciousness of creating a solution.

References

Billingham, Peter. *Edward Bond: A Critical Study*. London: Palgrave Macmillan, 2014.

---

[66] Billingham, *Edward Bond*, 5-6.
[67] Reinelt, *After Brecht*, 60.
[68] Anthony Jenkins, "Edward Bond: A Political Education," in *British and Irish Drama since 1960*, ed. James Acheson (Great Britain: Macmillan Press, 1993), 114.
[69] Bond, *The War Plays*, III.i.51.
[70] Witham, "English Playwrights," 300.
[71] Daneshzadeh, "Analysis of Edward Bond's *War Plays*," 2.

Booker, Keith M. *The Dystopian Impulse in Modern Literature: Fiction as Social Criticism*. USA: Greenwood Press, 1994.

Bond, Edward. *The War Plays: A Trilogy*. Great Britain: Methuen, 1985.

Carter, Ronald, and John McRae. *The Routledge History of Literature in English: Britain and Ireland*. 2nd ed. New York: Routledge, 2001.

Castillo, Debra A. "Dehumanized or Inhuman: Doubles in Edward Bond." *South Central Review*, no. 2. (1986): 78-89. http://www.jstor.org/stable/3189368.

Daneshzadeh, Amir. "Analysis of Edward Bond's *War Plays*." In *International Letters of Social and Humanistic Sciences*, 1-6. Switzerland: SciPress Ltd, 2015.

Hostetter, Robert D. "'Drama of the Nuclear Age': Resources and Responsibilities in Theatre Education." *Performing Arts Journal*, no. 2 (1988): 85-95. http://www.jstor.org/stable/3245646.

Innes, Christopher. "The Political Spectrum of Edward Bond: From Rationalism to Rhapsody." In *Contemporary British Drama, 1970-90: Essay from Modern Drama*, edited by Hersh Zeifman and Cynthia Zimmerman, 81-100. London: Macmillan, 1993.

Jenkins, Anthony. "Edward Bond: A Political Education." In *British and Irish Drama since 1960*, edited by James Acheson, 103-16. Great Britain: Macmillan Press, 1993.

Jones, Daniel R. "Edward Bond's 'Rational Theatre'." *Theatre Journal*, no. 4 (1980): 505-17. http://www.jstor.org/stable/3207412.

Lamont, Rosette Clémentine. "Edward Bond's DE-LEAR-IUM." *The Massachusetts Review*, no. 1/2 A Gathering in Honor of Jules Chametzky (2003): 308-13. http://www.jstor.org/stable/25091942.

Patterson, Michael. *Strategies of Political Theatre: Post-war British Playwrights*. Great Britain: Cambridge UP, 2003.

Rabey, David Ian. *English Drama since 1940*. Great Britain: Pearson Education, 2003.

Reinelt, Janelle. *After Brecht: British Epic Theater*. USA: The University of Michigan Press, 1996.

Spencer, Jenny S. *Dramatic Strategies in the Plays of Edward Bond*. Great Britain: Cambridge UP, 1992.

Stoll, Karl-Heinz. "Interviews with Edward Bond and Arnold Wesker". *Twentieth Century Literature*. No. 4. (1976): 411-32. http://www.jstor.org/stable/440583.

Witham, Barry B. "English Playwrights and the Bomb." *International Journal of Politics, Culture, and Society*, no. 2 (1988): 287-301. http://www.jstor.org/stable/20006902.

Chapter Six

Transforming Bodies in Maggie Gee's *The Ice People* and HBO Max's *Raised by Wolves*

Niğmet Çetiner

Introduction

We are now living in an age of artificial intelligence, which may appear to be a valuable development offering a friendly helping hand in dealing with everyday chores at home or allowing robots to function like human workers in factories, saving human beings as much time and money as possible. Nevertheless, human beings fail to recognize the "matter" value and the subsequent ethical concerns regarding new robot "species" as part of the nonhuman world. As human beings continue to develop robot technology, they appear to rely on human grandeur and ignore the imminent danger they unknowingly unleash on the world along with the rise of technology. While constantly demanding more advanced technology, humans have failed to see that artificial intelligence has continued to develop and indeed evolved into something far beyond what was originally intended. This raises the possible danger of the collapse of the family institution, cultures, civilizations, and the human world as we know it. At the same time, it lays the basis for a posthuman world with humans heavily interconnected with technology and robots evolving into human-like bodies. With this in mind, Maggie Gee's climate fiction, *The Ice People*, presents highly developed robots, the Doves, created for the pure pleasure and benefit of human beings in a posthuman society fighting for survival in an anthropogenic world, whereas HBO Max's TV series depicting a postapocalyptic world, *Raised by Wolves*,[1] introduces a robot character,

---

[1] *Raised by Wolves* was directed by Ridley Scott, Luke Scott, Sergio Mimica-Gezzan, Alex Gabassi, and James Hawes and aired in September 2020 on HBO Max.

Mother, who is originally created as a deadly weapon but later reprogrammed by an atheist hacker to be the parent of children who are intended to be the first members of an atheist colony to be founded on the planet Kepler-22b. The robots in both the book and the TV series manage to evolve by themselves and adopt human qualities, posing a great threat for human societies. Hence, as cli-fi works, *The Ice People* and *Raised by Wolves* demonstrate the dangerous outcomes of human arrogance failing to see the intricate value of technology in the posthuman world.

Posthumanism is a theory that undertakes the task of redefining the ontology of the human by challenging the Western mode of thinking that places humans above the nonhuman as supreme beings with reason. Therefore, it "designates a series of breaks with foundational assumptions of modern Western culture: in particular, a new way of understanding the human subject in relationship to the natural world in general."[2] Cary Wolfe presents the origin of posthumanism in the introduction of his book entitled *What is Posthumanism?* According to him:

> …posthumanism may be traced to the Macy conferences on cybernetics from 1946 to 1953 and the invention of systems theory involving Gregory Bateson, Warren McCulloch, Norbert Wiener, John von Neumann, and many other figures from a range of fields who converged on a new theoretical model for biological, mechanical, and communicational processes that removed the human and *Homo sapiens* from any particularly privileged position in relation to matters of meaning, information, and cognition.[3]

As is understood from this excerpt taken from Wolfe's book, posthumanism rejects the notions of traditional Western humanism, which puts humans at the center, claiming that they have a privileged position compared to nonhumans as the only beings with independence and cognition, and it aims to decenter humans; therefore, in contrast to Renaissance humanism, it is not an anthropocentric theory. It strives to deconstruct the Cartesian dualism that privileges reason over the corporeal body and seeks a flat ontology.[4] Therefore, "[posthumanism] seeks to undermine the

---

[2] David Jay Bolter, "Posthumanism." In *The International Encyclopedia of Communication and Philosophy*, eds. Klaus Bruhn Jensen and Robert T. Craig (United Kingdom: Wiley Blackwell, 2019), 1556-1559.
[3] Cary Wolfe, *What is Posthumanism?* (Minneapolis: The University of Minnesota Press, 2010), xi-xxxiv.
[4] Kübra Baysal, "Posthümanist Eleştiri ve Steven Hall'ün Köpekbalığı Metinleri Romanı." In *Yeni Tarihselcilikten Posthümanist Eleştiriye Edebiyat Kuramları*, ed. Mehmet Akif Balkaya (Konya: Çizgi Yayınevi, 2020), 209-233.

traditional boundaries between the human, the animal, and the technological."[5] As Michel Foucault asserts, "man is an invention of the recent date. And perhaps nearing its end,"[6] and humans will realize at last that they are not superior to nonhumans and they are governed by cosmic forces and nature rather than being the "masters and processors of nature."[7] In addition, they are evolving with the help of technology as much as they contribute to its evolution. Accordingly, Gee's *The Ice People* and the TV series *Raised by Wolves* provide a medium to illustrate posthumanist notions as there are humanoid robots that blur the boundary between the human and the technological with "artificial agents [becoming] capable of autonomous decisions,"[8] proving that humans are not the only beings with agency and the capacity to act independently.

In the same vein, it will be useful to refer to Donna Haraway's definition of cyborgs before delving into the universes of *The Ice People* and *Raised by Wolves*, both of which are teeming with humans sharing their worlds with the products of technoscience and needing technology to survive and evolve. Donna Haraway defines a cyborg as a cybernetic organism that is a hybrid of machine and organism in "A Cyborg Manifesto."[9] Pieces of fiction such as the *Terminator* movies (1984-1991) or *Doctor Who* television series (2005) are replete with cyborgs according to Haraway's definition. Additionally, as Haraway asserts, a cyborg is a creature of social realism, as well. Humans living in the contemporary world may also be regarded as cyborgs because there have been profound developments in biological sciences that enable genetic modification and xenotransplantation,[10] blurring the difference between human and nonhuman animals. In this regard, the book and the TV series explored here

---

[5] Bolter, "Posthumanism," 1556.
[6] Michel Foucault, *The Order of Things: An Archaeology of the Human Sciences* (United Kingdom: Taylor and Francis, 2005), 422.
[7] René Descartes, *A Discourse on the Method of Correctly Conducting One's Reason and Seeking Truth in the Sciences.* Trans. Ian Maclean (United Kingdom: Oxford University Press, 2006), 51.
[8] Michael Adrian Peters and Petar Jandrić, "Posthumanism, Open Ontologies and Bio-digital Becoming: Response to Luciano Floridi's Online Manifesto," *Educational Philosophy and Theory* (Jan 2019): 971-980.
[9] Donna Haraway, "A Cyborg Manifesto." In *Cultural Theory: An Anthology*, eds. Imre Szeman and Timothy Kaposy (United Kingdom: Wiley Blackwell, 2011), 454-471.
[10] Franklin Ginn, "Posthumanism." In *The International Encyclopedia of Geography*, eds. Douglas Richardson, Noel Castree, Michael F. Goodchild, Audrey Kobayashi, Weidong Liu, and Richard Marston (United Kingdom: Wiley and Sons, 2017), 5269-5277.

epitomize this point with humanoid robots that evolve and gain more and more anthropomorphic features and with humans that are dependent on these robots as well as other pieces of technology.

Maggie Gee's *The Ice People* and HBO Max's *Raised by Wolves* from the Perspective of Posthumanism

Maggie Gee's *The Ice People* (1998) is a climate dystopia presenting a future world first heated as a consequence of anthropogenic climate change and then cooled, resulting in an approaching new ice age. It describes societal decline with the contributions of science and technology going hand in hand with environmental collapse. The story centers around the nuclear family of Saul, Sarah, and their long-desired son Luke. The story is told in first-person narrative by Saul. Saul narrates his life story as he writes it down for wild children to read; his survival among these wild children hangs in the balance because, once the children tire of his stories, they will kill him. The novel opens with Saul, the narrator and protagonist, who was born in 2005 and is over sixty now, telling the story of his life to the wild children, a group of outlaws who have run away from their parents to live a Paleolithic lifestyle. As Saul proceeds, it is revealed that once, when the days were hot, there were many human beings, which is the first indicator of the dramatic population decrease. From the very beginning of the novel, a number of changes are indicated. People used to live twice long as Saul, who is over sixty, and if they were rich and lucky enough, they would not be "terminated"[11] (it is gathered later that, as can also be understood from the meaning of the word itself, being "terminated", or "termed" for short, denotes being euthanized). Saul's mother was a care assistant in an institution called the Last Farewell Home, where people over one hundred were terminated if they had no family members to take care of them. Saul's parents wanted to use the "terming" program, refusing to replace their organs, which would have enabled them to have a longer life. The old who were rich enough and had their organs replaced were referred to by the nickname "bits" by the young.[12] In the days when Earth was warming, water was scarce and watering a garden with tap water was a crime. Furthermore, global climate change led to a refugee crisis. The people who formerly resided in the now hotter countries got into or tried to get into England. This, in turn, culminated in othering and terming them as

---

[11] Maggie Gee, *The Ice People* (London: Telegram, 2012), chap. 1, www.telegrambooks.com.
[12] Ibid., chap. 6.

"Outsiders." Racism toward black people heightened with the onset of the Tropical Time in England, but all this changes with glaciation through the course of the novel. With the shrinking human population and the refugee crisis, civil order breaks down, although life goes on as if nothing unfavorable has happened in the wealthy areas of London. Well-off people who can afford to live in smart buildings, built by corporations such as StartSmart Buildings Inc., are protected from criminals and stay cool as anthropogenic climate change makes the rising temperatures unbearable. In addition, people experience a kind of cultural evolution. The way men and women dress, their hairstyles, and even their choice of partners change and homosexuality prevails, contributing to the already dwindling numbers of humans. In the meantime, science and technology reach their peak. For example, a great proportion of the society in England is nourished by consuming pills and Fibamix. Additionally, couples who cannot have children naturally resort to help from science, which leads to deformities in their offspring. In the same vein, it is suggested that the advancements in the field of genetics have interfered with the nature of all nonhuman beings, as well, as apparent from Saul's remark: "I tried to make her see that now nothing was natural, that the flowers she loved had been selectively bred to make them bigger and longer lasting, that even the hills behind the Northwest Borders, which we could just glimpse from our fourth floor window, were covered with genetically modified crops..."[13] Time and again, this change is revealed through smart buildings and minicopters, as well as the domestic robots or "mobots" called Doves, which are introduced as the story unfolds.[14] The Doves (the name stands for "DO VEry Simple things") are "humanoid robots [that] are becoming increasingly more human-like"[15] by evolving with or without human intervention and adopting human characteristics such as having simple conversations with humans, as understood from Saul's remark about a Dove named Dora. He even feels the obligation to explain to Dora why she cannot come to Africa with him and Luke when they decide to migrate. Dora, acting as if she understands human emotions and empathizes with them, asks Saul if he is unhappy, which shocks him: "'You are very unhappy?' ... 'I'm sorry that you're unhappy.' ... 'I feel. Very bad and very sad,' she said, shocking me to my

---

[13] Ibid., chap. 6.
[14] Ibid., chap. 6.
[15] Gönül Bakay, "Literary visions of post-apocalyptic worlds in the works of Mary Shelley, Margaret Atwood and Maggie Gee," *Esboços: Histórias em Contextos Globais* 27, no. 46 (December 2020): 220. doi: https://doi.org/10.5007/2175-7976.2020.e735 78.

selfish centre."[16] The initial purpose of the Doves when they were created was to clean, clear rubbish, walk, and talk, but later they are provided with other capabilities such as keeping people warm (Warmbots) and secure (Hawks), as well as appeasing their carnal desires (Sexbots). However, it is revealed as the story unfolds that they have various other faculties, which makes the Doves central to this study's argument.

Saul, the narrator and protagonist of the novel, was born in London in 2005, which corresponds to the beginning of the period marked by global warming, or "the Tropical Time,"[17] as it is called in the book. Technology brings about environmental collapse with anthropogenic global warming and subsequent glaciation. Moreover, fertility rates fall, not only because men and women choose to live separately, but also for biological reasons such as the sterility of women or low chances of sperm being able to fertilize eggs because of the unwanted effects of advanced science. When the Doves are introduced, the decline in the population and the segregation of men and women increase because men who seek comfort and consolation buy Doves to replace their wives and children, while women flock around children, using them as a medium through which to assert power over the opposite sex. In the novel, Saul assumes that even machines are endowed with agency. He loves "artificial life" and has dealt with robots for years, although he later turns down job offers related to his lifelong fascination and becomes a tech teacher in a Learning Centre:

> I found I had a gift with machines. They were alive to me, and entirely absorbing, like the aphids I once bred in a matchbox. I was fascinated by artificial life, by the huge range of mobots in the college labs, the multitravellers, the swarmers, the sorters, though my specialty was nanotechnics, working with invisibly small molecular machines. I had delicate powers of manipulation that helped me pass out with high honours. Job offers came in plenty from military and security firms. For some reason I found myself turning them down. My father was shocked, but I knew I wasn't ready. Something had to happen first, some great adventure. For the moment I took a part-time job as a tech teacher in a Learning Centre.[18]

He even greets the voicetone of the school building and states his belief in the agency of the artificial: "I always said 'Good Morning' back, though the teachers laughed at me. They thought I was joking but I wasn't.

---

[16] Gee, *The Ice People*, chap. 19.
[17] Ibid., chap. 2.
[18] Ibid., chap. 2.

It seemed to me anything might be alive."[19] In Saul's voice, Gee articulates a question that lies at the core of posthumanism: "What was the boundary between living and nonliving?"[20] In addition, Saul gives his Dove a human name, Dora, and treats it as if it were an individual. He hesitates to give her away to the strangers with whom he bargains for a boat. Saul's interactions with machines and his belief in their potential regarding their vitality, making him feel that they may have consciousness, are analogous to Nancy Katherine Hayles's views on "machine cognizers" in the contemporary world, where data flows occur between machines for the most part.[21] Humans are being immersed into these cognitive systems and they are not the only actors in the electromagnetic spectrum. Machines are also involved in these systems and they are evolving, becoming more and more cognitive, which is yet further proof refuting the human and machine divide.

> In highly developed and networked societies such as the US, human awareness comprises the tip of a huge pyramid of data flows, most of which occur between machines. Emphasizing the dynamic and interactive nature of these exchanges, Thomas Whalen (2000) has called this global phenomenon the cognisphere. Expanded to include not only the Internet but also networked and programmable systems that feed into it, including wired and wireless data flows across the electromagnetic spectrum, the cognisphere gives a name and shape to the globally interconnected cognitive systems in which humans are increasingly embedded. As the name implies, humans are not the only actors within this system; machine cognizers are crucial players as well. If our machines are 'lively' (as Haraway provocatively characterized them in the 'Manifesto'), they are also more intensely cognitive than ever before in human history.[22]

In contrast to Saul, the Wicca, a collective of women who worship nature and try to find ways to reproduce without male intervention, exclude men from all aspects of their life: "They didn't want [them] as lovers, or fathers, or friends."[23] As supporters of the natural, the women of the Wicca condemn everything artificial, including the Doves, and they do the same to men who have used Doves as substitutes for their sons, daughters, wives, and lovers. It could be inferred in light of this information that the

---

[19] Ibid., chap. 2.
[20] Ibid., chap. 2.
[21] Nancy Katherine Hayles, "Unfinished Work: From Cyborg to Cognisphere," *Theory, Culture & Society* 23, no. 7-8 (December 2006): 159-166. doi: https://doi.org/10.1177/0263276406069229.
[22] Ibid., 161.
[23] Gee, *The Ice People*, chap. 8.

constituents of nature and culture change sides here and men are disdained, being associated with the cultural and artificiality.

The Doves, on the other hand, prove Saul's point about mechanical beings having agency rights. The Doves are robots produced to perform household chores in place of humans. Initially, there was only one primitive model; later, with the help of nanotechnology, a few more models were produced. These models have more functions than the earliest Doves, such as the ability to provide pleasure or to self-replicate. Later, it is revealed that the offspring of the Doves mutate as they replicate. These mutants run away from homes, gather in flocks, and attack people. They show signs of sentience by escaping from homes and gathering together. Thus, humans lose control over the technology they create. This particular situation exemplifies Ihab Hassan's argument as presented in his article entitled "Prometheus as Performer: Toward a Posthumanist Culture?":

> Will artificial intelligences supersede the human brain, rectify it, or simply extend its powers? We do not know. But this we do know: artificial intelligences, from the humblest calculator to the most transcendent computer, help to transform the image of man, the concept of the human. They are agents of a new posthumanism, even if they do no more than IBM 360-196 which "performs in a few hours all the arithmetic estimated ever to have been done by hand by all mankind."[24]

While Gee's *The Ice People* is about human parents caring for and raising their own children on an Earth that is experiencing a global climate catastrophe and has been changed by technology, *Raised by Wolves* is about two androids trying to parent human children on an alien planet in alignment with the mission assigned to them by their creator. *Raised by Wolves* is a science fiction TV series created by Aaron Guzikowski (IMDB 2020). It started airing on September 3, 2020, on HBO Max. It is set in a distant future on a postapocalyptic Earth that became uninhabitable after being scourged by a war between the religious zealots of Sol, with a religious belief based on the sun god of a Mithraic cult, Sol Invictus of Ancient Rome,[25] and the Atheists. The story begins with the landing of two androids, Mother and Father, along with twelve frozen human embryos in a space vessel on

---

[24] Ihab Hassan, "Prometheus as Performer: Toward a Posthumanist Culture," *The Georgia Review* 31, no. 4 (Winter 1977): 830-850.
https://www.jstor.org/stable/41397536?seq=1#metadata_info_tab_contents.
[25] Juliette Harrison, "Raised by Wolves: Mithraism and Sol Explained," accessed January 2, 2021. https://www.denofgeek.com/tv/raised-by-wolves-mithraism-sol-explained/.

Kepler-22b. Mother and Father are tasked with initiating an Atheist colony on the planet by Campion, an Atheist hacker who captured and reprogrammed Mother, a former Mithraic android that was originally a necromancer. Mother nurtures the fetuses attached to her body through artificial umbilical cords and removes them from the artificial wombs when the time is right. Although she tries hard, she succeeds in raising only one of the babies, Campion, whom she names after her creator. As the story proceeds, a Mithraic spaceship appears in the atmosphere of the planet. Mother regards the Mithraic crew as dangerous as she thinks that they are going to take Campion away. She kidnaps five children from the ship and then crashes it with all the other children and adults inside. In due course, Mother and Father strive to proceed with their mission to set up an Atheist colony with the children despite continuous attacks from the survivors of the Mithraic spaceship.

It is possible to observe human to nonhuman together with nonhuman to human transformation in *Raised by Wolves*. This transformation calls for "a repositioning of the human among nonhuman actants"[26] as it challenges the culturally constructed boundary between humans and machines as well as between humans and animals. In addition, the "self and other" dichotomy wreaks havoc on nature and humanity. For instance, the Mithraic consider themselves superior to animals and androids, and even to their fellow humans, the Atheists. They believe that they are the only ones whose souls are pure and worth praying for. Concordantly, they do not pray for the souls of Atheists because their souls are considered impure. Moreover, according to the Mithraic doctrine, animals and androids have no souls to pray for at all. Therefore, they put the androids at their disposal and even send them off to die in their stead. The only beings imbued with subjectivity are humans; that is, according to the Mithraic, only humans have consciousness, agency, and personhood. However, androids disprove their belief time and again. To illustrate, when one of Ambrose's androids is attacked and thrown into a pit by Mother while trying to lure her away from the Mithraic, her sister wails for her at the edge of the pit, manifesting her sorrow.[27] Furthermore, Mother does not want to dispose of the dead baby Campion of the Gen-1s, the children whose frozen fetuses were brought to Kepler 22-b by Mother and Father. She refuses to let go of the stillborn fetus and grieves over it in the same way a human mother would.[28] She constantly questions her motherhood, and she suspects that she is the one that killed the Gen-1s and

---

[26] "New Materialism(s)," Criticalposhumanisms.net, accessed Jan. 2, 2020, https://criticalposthumanism.net/new-materialisms/.
[27] *Raised by Wolves*, "Virtual Faith."
[28] *Raised by Wolves*, "Raised by Wolves."

suffers for it. In her dialogue with Tempest, another child Mother abducted from the Mithraic ship, she indicates that she wants to have a baby of her own, which proves that she has started to think and act with free will and she does not remain as Campion programmed her. She gradually shows signs of becoming a sentient and subjective being making autonomous decisions: "Do you know how fortunate you are? I've always wanted a child that came from me. You are a creator, whereas all I'll ever be…is a creation."[29] In addition, Father assumes that he cannot protect the children as well as Mother does and he fears he will lose Mother's respect. He even shows signs of love and jealousy when Mother leaves the settlement frequently to connect to the simulator she found: "…your prolonged absences. Sometimes cause me. To cycle through. Various scenarios… Where I have to hypothesize the nature of your activities."[30] These events demonstrate signs of Mother and Father's gradual development of self-awareness and self-doubt, which proves that the idea that humans are the only beings endowed with faculties such as intelligence, cognition, insight, and emotion is a misconception.

Campion is the counterpart of Saul in *Raised by Wolves*. In a dialogue with Paul, another Mithraic child abducted by Mother, Campion articulates his thoughts about the fact that everything has a soul and should not be killed, including the native creatures[31] on Kepler-22b, which are later revealed to be devolved humans. Paul, as a child raised with Mithraic beliefs, shares the ideas of the Mithraic about animals and androids[32] having no souls even though he has a pet mouse. He tells Campion that as his father read it in the scriptures, it is not necessary to bury the bones of those native creatures; however, Campion disagrees with him: "Just because you say it doesn't make it true. … I think everything has a soul. Even Mother and Father. Maybe even trees. The big ones, anyway."[33] The Mithraic religion deems humans, especially the believers in the Mithraic religion, superior to nonhuman animals and inanimate beings. In a similar vein, it is written in the Gospels that humans are superior to animals.[34] Both religions, Mithraism and Christianity, value humans above the nonhuman, supporting

---

[29] *Raised by Wolves*, "Nature's Course."
[30] *Raised by Wolves*, "Lost Paradise."
[31] With the introduction of these creatures, it is revealed to the viewer that Kepler-22b is not a virgin planet; on the contrary, it has a past. In addition to these creatures, this fact is also made obvious by certain artifacts on the planet.
[32] *Raised by Wolves*, "Faces."
[33] *Raised by Wolves*, "Lost Paradise."
[34] Laura Hobgood-Oster, *Holy Dogs and Asses: Animals in the Christian Tradition* (Illinois: The Board of Trustees, 2008), 57.

the ideals of Western humanism. The Mithraic do not only discriminate between humans and nonhumans; they also categorize humans into two groups as believers and nonbelievers. Their fanaticism leading to natural/artificial, human/nonhuman, and believer/nonbeliever binaries drags life on Earth to its ultimate demise.

Both *The Ice People* and *Raised by Wolves* depict "a post-apocalyptic world, where the 'post' indicates a spatial-temporal situation that follows the destruction of the human world – with all its social, political and cultural structures – as we know it today."[35] In *The Ice People*, the birth rate dwindles as a result of humans' detrimental impact on nature and that leads to a dramatic decrease in the population and to a clash between genders. Animals become extinct and the initial rise in temperature is replaced by glaciation. Climate refugees appear, increasing the fear of violence. Humans resort to cannibalism because of the scarcity of food. In a similar manner, *Raised by Wolves* introduces a planet destroyed and made inhospitable by the war between the Atheists and the religious zealots of Mithras who escape the planet in their advanced spaceship, leaving animals, the powerless, and the Atheists, their enemies, behind. Both pieces of fiction state the obvious; human exceptionalism renders the nonhuman insignificant and exploitable, which leads to humanity's own downfall in turn. In both works, it is not the Doves or the androids that bring about ecological collapse; it is the anthropocentric hubris that blinds humans and tricks them into exploiting their environment, androids, and even fellow humans. Therefore, humans need to be decentered and they must rethink and redefine their place among nonhumans.

In addition, *Raised by Wolves* features androids that perform various tasks such as providing services ranging from healthcare to security. As can be gathered from these examples, it is not necessary to have a piece of machinery incorporated into a human body to be considered a cyborg. Humans' dependence on machines increased after the Industrial Revolution and machines became like extensions of the human body. However, according to Haraway, humans have always been cyborgs. They have always been evolving alongside technology. As maintained by Ihab Hassan, humans became posthuman with the control of fire. He draws on the myth of Prometheus while developing his view. By stealing fire from the gods and giving it to humans, Prometheus bestowed "knowledge and imagination, the alphabet, medicine, and all arts."[36] Anne Harrison explains

---

[35] Antonio Lucci, "Genealogies of Posthuman Narratives: Critical Theory, Animal Studies and the (Beyond of) Neolithic Revolution," *(Re) Pensar El Mundo. Nuevas Tendencias En El Pensamiento Contemporaneo* 3, no. 3 (December 2017): 435-459.
[36] Hassan, "Prometheus as Performer," 830-850.

the idea of fire as the thing that made the human posthuman by using today's knowledge:

> ... it is fire that allowed *Homo erectus* to sleep safely at night on the ground; it is fire that cooked the food that nourished and developed our outrageous brains; and it is fire that made us human. In our intertwined ontologies we collapse materiality and metaphor: we are fiery, and fires die. We can understand these fusions and confusions when we see the bricks of Babel remain as relics of the architecture of a common language; when we see glass mediate impossible truths of immanent matter; and when we grant that pyrophiliacs feel the objecthood and artistry of being made by fire. We can realize it: we are pyromena – we are fire's doing.[37]

The Ice People and *Raised by Wolves* introduce posthuman worlds in which humans depend on technology. They struggle to survive the hostile conditions that they created via the same technology to which they cling. Androids, spaceships, and advanced weapons in *Raised by Wolves* and smart buildings, minicopters, and the Doves in *The Ice People* are pieces of technology that help humans improve their conditions. The destruction brought upon humans by technology proves posthumanism's point about human exceptionalism. In other words, human exceptionalism grants humans the right to exploit the environment, leading to its destruction. In a similar vein, traces of transhumanism are also observable in the form of promoting science and the celebration of human progress. In *The Ice People*, people whose organs start to fail because of old age replace them and thus live a long life, improving their life quality. In addition, in *Raised by Wolves*, Ambrose, the Mithraic leader, has an android hand. In both pieces of fiction, science and technology are both helpful and harmful. Technology helps to improve human lives but, conversely, it destroys both humans and nonhumans when misused.

As Francesca Ferrando argues, an approach that does not prioritize one entity over another for any reason is needed, and posthumanism provides such an approach. It strives to refute the conviction that holds humans as supreme beings, to equalize their ontology with that of nonhumans, and to take a stand against anthropocentrism, speciesism, or biocentrism. It underlines human and nonhuman interdependence. According to Ferrando, a society with such an awareness is a posthuman

---

[37] Anne Harris, "Pyromena *Fire's Doing*." In *Elemental Ecocriticism: Thinking with Earth, Air, Water, Fire*, ed. Jeffrey Jerome Cohen and Lowell Duckert (Minneapolis: University of Minnesota Press, 2015), 47-48.

society that embraces symbiosis with the nonhuman and never assumes a speciesist attitude:

> Existential dignity invites an approach that is not hierarchical nor based on socio-political, biological or planetary, supremacies. More clearly, Posthumanism does not support the relativization of human, and non-human, life. The goal is not to perpetuate human discriminations (in its various forms), nor to replace them with other forms of discriminations, such as anthropocentrism, speciesism or biocentrism, among others. Sustained by this awareness, we can eventually achieve a posthuman society based on co-existence and multi-species justice.[38]

However, in parallel with the posthuman notion deliberated by Ferrando, the book and the TV series reflect on various kinds of dualities that are byproducts of Enlightenment humanism. In *The Ice People*, people suffer the devastating results of the inferiorization of nature and natural others, facing anthropogenic climate change and the subsequent ice age. The book reflects on the dichotomy of nature and culture, providing ecological insight into the inseparability, interdependency, and equality of the natural and cultural realms and the fact that the cultural needs the natural in order to survive. Furthermore, the racist attitude toward the Outsiders, or non-white people, dissolves when the white become climate refugees striving to take shelter in Africa. Moreover, *Raised by Wolves* presents two worlds, one scourged by violence between two warring factions, the Mithraic and Atheists, that deem themselves superior to each other and also to the nonhuman, while the other is hinted to have suffered the same fate at some point in the past. In both works, the disregard of humanity's dependence on the nonhuman leads to catastrophic consequences such as extinctions, including the near extinction of the human species itself.

Conclusion

In both *The Ice People* and *Raised by Wolves*, humans are intertwined with nonhumans, both shaping and shaped by them. In both stories, parts of the human body can be replaced with new ones thanks to technology. The products of human technology provide various services from cleaning and providing sexual comfort to raising human children or building a civilization that values science over religion. Moreover, the

---

[38] Francesca Ferrando, "Leveling the Posthuman Playing Field," *Theology and Science* 18, no. 1 (February 2020): 1-6. doi: https://doi.org/10.1080/14746700.2019.1710343.

machines in these works transform into human-like beings by evolving and gaining anthropomorphic features while android organs replace human body parts. Therefore, analyzing these two works that belong to two different genres with regard to posthumanism facilitates the questioning of what it means to be human and the place of humans among nonhumans.

References

Bakay, Gönül. "Literary visions of post-apocalyptic worlds in the works of Mary Shelley, Margaret Atwood and Maggie Gee." *Esboços: histórias em contextos globais* 27, no. 46 (December 2020): 405-425. https://doi.org/10.5007/2175-7976.2020.e735 78.

Baysal, Kübra. "Posthümanist Eleştiri ve Steven Hall'ün Köpekbalığı Metinleri Romanı". In *Yeni Tarihselcilikten Posthümanist Eleştiriye Edebiyat Kuramları*, edited by Mehmet Akif Balkaya, 209-233. (Konya: Çizgi Yayınevi, 2020).

Bolter, Jay David. "Posthumanism." In *The International Encyclopedia of Communication and Philosophy*, edited by Klaus Bruhn Jensen and Robert T. Craig, 1556-1559. (UK: Wiley Blackwell, 2016).

Descartes, René, *A Discourse on the Method of Correctly Conducting One's Reason and Seeking Truth in the Sciences*. Translated by Ian Maclean. Oxford: Oxford University Press, 2006.

Ferrando, Francesca. "Leveling the Posthuman Playing Field." *Theology and Science* 18, no. 1 (February 2020): 1-6. https://doi.org/10.1080/14746700.2019.1710343.

Foucault, Michel. 2005. *The Order of Things: An archaeology of the Human Sciences*. UK: Wiley Blackwell.

Gabassi, Alex, dir. *Raised by Wolves*. Season 1, Episode 7, "Faces." Aired Sep 17, 2020, on HBO Max.

Gee, Maggie. *The Ice People*. London: Telegram, 2012.

Ginn, Franklin. "Posthumanism". In *The International Encyclopedia of Geography*, edited by Douglas Richardson, Noel Castree, Michael F. Goodchild, Audrey Kobayashi, Weidong Liu and Richard Marston, 5269-5277. (UK: Wiley & Sons, 2017).

Harris, Anne. "Pyromena Fire's Doing". In *Elemental Ecocriticism: Thinking with Earth, Air, Water, and Fire*, edited by Jeffrey Jerome Cohen and Lowell Duckert, 27-54. (Minneapolis: University of Minnesota Press, 2015).

Harrison, Juliette. "Raised by Wolves: Mithraism and Sol Explained." Accessed Jan 2, 2021. https://www.denofgeek.com /tv/raised- by-wolves-mithraism-sol-explained/.

Hassan, Ihab. "Prometheus as Performer: Toward a Posthumanist Culture?" *The Georgia Review* 31, no. 4 (Winter 1977): 830-850.

Hayles, Katherine. "Unfinished Work From Cyborg to Cognisphere." *Theory, Culture & Society* 23, no. 7-8 (December 2006):159-166. https://doi.org/10.1177/0263276406069229.

Hobgood-Oster, Laura. 2008. *Holy Dogs and Asses: Animals in the Christian Tradition*. Illinois: The Board of Trustees.

Lucci, Antonio. "Genealogies of Posthuman Narratives: Critical Theory, Animal Studies and the (Beyond of) Neolithic Revolution." *(Re) Pensar El Mundo. Nuevas Tendencias En El Pensamiento Contemporaneo* 4, no. 3 (Decemeber 2017): 435-459.

Peters, Michael and Petar Jandrić. "Posthumanism, Open Ontologies and Bio-digital becoming: Response to Luciano Floridi's Onlife Manifesto." *Educational Philosophy and Theory* 51, no. 10 (2019): 971-980. https://doi.org/10.1080/00131857.2018.1551835.

Mimica-Gezzan, Sergio, dir. *Raised by Wolves*. Season 1, Episode 6, "Lost Paradise." Aired Sep 17, 2020, on HBO Max.

Raised by Wolves, directed by Ridley Scott, Luke Scott, Sergio Mimica-Gezzan, Alex Gabassi, James Hawes, aired in September, 2020 on HBO Max.

Sanzo, Kameron. "New Materialism(s)." Accessed January 2, 2020. https://criticalposthumanism.net/new- materialisms/.

Scott, Luke, dir. *Raised by Wolves*. Season 1, Episode 3, "Virtual Faith." Aired Sep 3, 2020, on HBO Max.

Scott, Luke, dir. *Raised by Wolves*. Season 1, Episode 4, "Nature's Course." Aired Sep 10, 2020, on HBO Max.

Scott, Ridley, dir. *Raised by Wolves*. Season 1, Episode 1, "Raised by Wolves." Aired Sep 3, 2020, on HBO Max.

Wolfe, Cary. 2010. *What is Posthumanism?* Minneapolis: The University of Minnesota Press.

Chapter Seven

Society Islands and the Anthropocene in David Mitchell's *Cloud Atlas*

Emily Arvay

In the Malvern hills of Worcestershire, David Mitchell's "earliest creative acts" entailed filling "little notebooks" with maps of "imaginary archipelagos," which he named and peopled before contemplating their "international relations [and] wars" – childhood keepsakes Mitchell now considers to be his "first novels."[1] These early notebooks are all the more striking given that Mitchell's *magnum opus* is similarly preoccupied with island communities beset by war. Mitchell's *Cloud Atlas* (2004) mentions no less than five islands, and two of the novel's six narratives unfold on archipelagos: one historical (Rēkohu) and one typically understood to be speculative (Ha-Why). In keeping with Mitchell's "little notebooks," *Cloud Atlas* evinces the author's longstanding fascination with international relations, cultural predation, and war. The novel's distant past ("The Pacific Journal of Adam Ewing") unfolds on 1850s Rēkohu fifteen years after the Maori massacre of the Moriori, whereas the novel's distant future ("Sloosha's Crossin' an' Ev'rythin' After") unfolds on thirtieth-century Ha-Why during the Kona massacre of Big I's windward communities.[2] While literary critics rightly note the thematic congruencies that link these two besieged islands, most overlook how double-mapping devices give Hawai'i's recent past the appearance of a distant future. This chapter therefore contributes to contemporary scholarship by contending that Mitchell models Ha-Why's violent territorial clashes after a four-year period (1790-1794) that saw Kona forces under Kamehameha's command seize Hawai'i's Big Island. To the extent that *Cloud Atlas* diverges from Hawai'i's recorded history, the novel does so to foreshorten the historical gap that separates a distant past destabilized by intertribal warfare (1780s)

---

[1] Sam Leith, "A Literary Houdini," *Telegraph* (February 24, 2004), n.p.
[2] David Mitchell, *Cloud Atlas*, (London: Sceptre, 2004), 3-39; 493-529; 249-325.

from a more recent past (1790s) rocked by British colonial incursion.

Although Mitchell's novel makes no explicit mention of climate change, this chapter builds on Astrid Bracke's classification of *Cloud Atlas* as a text "with no obvious environmental dimension" that, nevertheless, lends itself to ecocritical interpretations.[3] Indeed, Mitchell's efforts to rewrite historical Hawai'i's eighteenth-century unification as a speculative future are in keeping with other climate fictions produced in the early 2000s. The release of a third assessment report (2001) by the Intergovernmental Panel on Climate Change (IPCC) heightened public urgency and ushered the term Anthropocene into common usage. As Adam Trexler notes in his expansive survey of Anthropocene fictions, the confluence of environmental, financial, and political crises that coalesced in the early 2000s not only altered the temporal horizon for human survival but also the parameters of British prose. To articulate the temporal and spatial reach of the climate crisis, British authors increasingly sought an experimental form capable of connecting events dispersed across vast historical timescales and geophysical distances. Notably, British novelists Charles Avery, Will Self, Jeanette Winterson, and David Mitchell all experimented with double-mapping devices to concretize the transhistorical impacts of contemporary climate change both locally and globally.[4] In particular, British climate fictions penned during Prime Minister Tony Blair's turbulent second term frequently deployed the "ruined-island-as-future-Earth" motif as a potent metonymic equivalent to transpose the harrowing histories of actual islands onto speculative archipelagos in order to render transhistorical crises contiguous with present-day catastrophe *and* future extinction.

Concurrent with the emergence of double-mapped climate fictions, the early 2000s also witnessed the proliferation of cultural studies and popular science texts[5] that sought to probe the "precise laws and specific effects" of topographical variables on global culture systems to extract transhistorical patterns with predicative applications.[6] Climate science journals, in turn, featured the climate-collapse histories of islanded communities on the pretext that such accounts might offer contemporary

---

[3] Astrid Bracke, "The Contemporary English Novel and its Challenges to Ecocriticism," in *Oxford Handbook of Ecocriticism*, ed. Greg Garrard (Oxford: Oxford University Press, 2014), 429.
[4] See Charles Avery's *The Islanders* (2004), Will Self's *The Book of Dave* (2006), and Jeanette Winterson's *The Stone Gods* (2007).
[5] See, for example, Michael Hardt and Antonio Negri's *Empire* (2001) or Ronald Wright's *A Short History of Progress* (2004).
[6] Patrick O'Donnell, "Introduction," in *A Temporary Future: The Fiction of David Mitchell*, ed. Patrick O'Donnell (London, United Kingdom: Bloomsbury, 2015), 12.

readers salutary lessons for correcting the short-sighted dependencies presumed to cement the historical demise of circumscribed populations.[7] Foremost among the climate-collapse genre was Jared Diamond's bestselling *Guns, Germs, and Steel* (1997) – a popular science text Mitchell lauded for inspiring thematic contents of *Cloud Atlas*. Apparently, it was Diamond's contention that "there is nothing inevitable about civilization" that first piqued Mitchell's interest in "society islands" and later propelled the author's decision to conduct field work on Rēkohu and Hawai'i.[8] While numerous critics concur that *Guns, Germs, and Steel* supplies much of the intellectual groundwork for the "Pacific Journal" portions of *Cloud Atlas*, critics have yet to extend this observation to the contents of "Sloosha's Crossin'" – a speculative future strewn with historical referents that predate the period portrayed in "Pacific Journal" by more than sixty years. As such, this chapter contends that "Sloosha's Crossin'" engages with the "Polynesian experiment" undertaken in *Guns, Germs, and Steel*, in which Diamond cites Kamehameha's unification of the Hawaiian archipelago as evidence that human predation constitutes a latent propensity enabled by enviro-materialist precondition.[9] Akin to Diamond, Mitchell's novel attributes human predation to the emergence of grain-fed kleptocracies sustained by food surpluses, ceremonial works, and territorial conquests. Yet, Mitchell's *Cloud Atlas* posits a more nuanced understanding of the relationship between environmental precarity, epistemological polarization, and communal scapegoating than Diamond. Indeed, Mitchell's novel depicts human predation not as a latency enabled by enviro-materialist variables but as a genre of thought enabled by intergenerational fictions of human progress. This chapter ultimately argues that Mitchell's double-mapped Ha-Why articulates a vision of human "essence" that is intergenerationally reconstituted through narratological acts of reception and transmission.

As Mitchell's "most ambitious experiment in narrative form" to date,[10] critics often spatialize the novel's chiastic structure by way of analogy. Thus far, critics have compared *Cloud Atlas'* nested form to a palindrome, an Ouroboros, a Chinese-box, and a Matryoshka doll. When interviewed, Mitchell likens his novel to "a row of ever-bigger fish eating

---

[7] David Correia, "F**k Jared Diamond." *Capitalism, Nature, Socialism* 24, no. 4 (2013): 4.
[8] Adam Begley, "The Art of Fiction No. 204," *Paris Review* 193 (Summer 2010): n.p.
[9] Jared Diamond, *Guns, Germs, and Steel* (New York: W.W. Norton & Co., 1997), 276-278.
[10] Courtney Hopf, "The Stories We Tell," in *David Mitchell: Critical Essays*, ed. Sarah Dillon (Canterbury, United Kingdom: Gylphi, 2011), 108.

the one in front.'"[11] In effect, the reader progresses from the earliest embedded narrative through a series of successive futures before reversing course in pursuit of an increasingly distant past. With the notable exception of "Sloosha's Crossin,'" the protagonist of each section recovers, interprets, and completes the narrative of their predecessor – thus restoring coherence to a previously fragmented text whilst generating a soon-to-be fragmented text of their own.[12] The structural placement of "Sloosha's Crossin'," as the outmost container for a series of embedded remediations, thus serves as the "fulcrum" around which the rest of the novel turns.[13] While *Cloud Atlas* has garnered a great deal of attention, praise, and critical scholarship since its debut (2004) and subsequent adaptation into a Hollywood film (2012), few literary scholars discuss "Sloosha's Crossin'" in detail as a discrete section worthy of critical attention. Indeed, most make brief mention of "Sloosha's Crossin'" as a means to cite Zachry's poignant comparison of souls to clouds from which the novel's title derives.[14] As the central hinge through which the novel's sections are recovered and completed in reverse, and as the only section relayed in its entirety without generating a text of its own, critics often describe "Sloosha's Crossin'" as the node through which Mitchell's many worlds are yoked.[15] Likewise, scholars often note the thematic recurrence of Hawai'i within the novel, which first surfaces as Autua's nineteenth-century destination, then resurfaces as a twentieth-century "radio-astronomy" site, thirtieth-century human slaughterhouse, and, finally, as the outermost diegetic world through which the novel's other worlds are accessed.[16] To the extent that critics attend to Mitchell's Ha-Why, most cite the thematic congruence of "Sloosha's Crossin'" to Ewing's "Pacific Journal" either to affirm the recursive fall of human civilization, or to underscore the illusion of total circumscription on which the will to power relies. In short, scholars have yet to account for the partially submerged referents that inform "Sloosha's Crossin,'" in which the "treble" of Zachry's

---

[11] Paul Ferguson, "'Me Eatee Him Up'," *Green Letters* 19, no.2 (2015): 146.
[12] Stephen Abell, "How to Get the Ker-Bam," *Times Literary Supplement* 5265 (February 27, 2004): 21.
[13] Jo Alyson Parker, "David Mitchell's *Cloud Atlas* of Narrative Constraints and Environmental Limits," in *Time: Limits and Constraints*, eds. Jo Alyson Parker, Paul Harris, and Christian Steineck (Boston: Brill, 2010), 201.
[14] Mitchell, *Cloud*, 324.
[15] Oliver Lindner, "Postmodernism and Dystopia," in *Dystopia, Science Fiction, Post-Apocalypse*, edited by Eckart Voights and Alessandra Boller (Trier: Wissenschaftlicher: Verlag Trier, 2015), 366; O'Donnell, "Introduction," 77-78.
[16] Mitchell, *Cloud*, 36; 96; 189-90; 249-325.

tale serves to complement history's "bass."[17] Instead, critics are almost unanimous in describing "Sloosha's Crossin'" as set in a distant future that returns humanity to a speculative stone age littered with technological remnants of the reader's own.[18] Consequently, critics are more inclined to read Mitchell's Kona in terms of Tolkien's "orcs" or the sci-fi villains of "Mad Max 3" than as a defamiliarized portrait of eighteenth-century Hawaiians.[19] Given this literary context, scholar Patrick O'Donnell stands apart for conducting a rigorous analysis of "Sloosha's Crossin'" to argue that the section exemplifies Mitchell's ambition to expose the narrative of Empire as dependent on historical remediation.[20] By extension, this chapter intervenes in extant criticism by contending that Mitchell's *Cloud Atlas* epitomizes a climate-fictional subgenre accessible to the reader only through extratextual engagements with history. To discern historical referents mapped onto speculative topographies extrapolated from the prognostications of the IPCC, the reader must connect the novel's semi-veiled allusions to those keyed into Mitchell's extant corpus. Indeed, to accommodate engagement with historical allusions woven across multiple texts, the reader must conceive of *Cloud Atlas* as David Mitchell does – as a discrete chapter within a "sprawling macro-novel."[21] This chapter thus foregrounds the historical dimensions of "Sloosha's Crossin'" to suggest that the reader cannot grasp the significance of Mitchell's *magnum opus* without first attending to the historical referents that undergird its seemingly speculative future.

This chapter proceeds by arguing that the parallels that Mitchell draws between 1850s Rēkohu and a fabular 1790s Hawai'i are not random but register the profound cultural impact of *Guns, Germs, and Steel*, in which Diamond performs a comparative analysis of Polynesian populations descended from a common ancestor to assess the impact of environmental variables on human predation. More specifically, "Sloosha's Crossin'" engages with the "Polynesian experiment" undertaken in the second chapter of *Guns, Germs, and Steel*, in which Diamond cites Kamehameha's unification of the Hawaiian archipelago as evidence that human predation

---

[17] David Mitchell, "On Historical," in *The Thousand Autumns of Jacob de Zoet* (London: Sceptre, 2011), 558.
[18] Sandrine Sorlin, "A Linguistic Approach to David Mitchell's Science-Fiction Stories in *Cloud Atlas*," *Miscelánea* 37 (2008): 76.
[19] Julie Morère, "*Cloud Atlas*," *PU de la Méditerranée* (2010): 290; Leith, "Literary," n.p.
[20] O'Donnell, "Introduction," 70.
[21] Wyatt Mason, "David Mitchell the Experimentalist," *New York Times Magazine* (June 25, 2010): n.p.

remains a latent propensity enabled by enviro-materialist precondition.²² Central to Diamond's theorization is the bifurcation of islanded communities into nomadic foragers and grain-fed colonizers, in which the former inevitably fall prey to the latter.²³ Using remote islands as "control sites" from which to test the impact of environmental variables, Diamond deems foraging communities prone to decentred egalitarianism and agricultural communities to expansionist predation, as the lee and windward communities of eighteenth-century Hawai'i demonstrate.²⁴ Although Diamond does not cast human predation as an innate biological drive, much of *Guns, Germs, and Steel* relies on the operative assumption that agricultural gluts enabled by climatological and topographical variables midwife a latent propensity towards ecocidal territorial expansion. As Diamond explains, prior to Cook's initial contact with "Owhyee" (1778), the Hawaiians enjoyed roughly two thousand years of cultural isolation.²⁵ By the time Cook's *Resolution* anchored in Kealakelua Bay (1779), inter-island "political fusion" was already underway.²⁶ Climatological propensities gifted the Hawaiians with permanent streams and arable soil.²⁷ Moreover, advanced agricultural techniques generated food surpluses sufficient to sustain complex political hierarchies, advanced technologies, and the construction of ceremonial temples – demonstrations of power that, over time, came to exceed the bounds of the archipelago's resources.²⁸ To obtain the materials necessary for continued temple construction, Hawai'i's hereditary chiefs forcibly expanded their territorial holdings. Throughout the 1780s, the chiefs of Hawai'i's four largest islands (Big Island, Maui, O'ahu, and Kaua'i) jockeyed for control over its four smallest (Lana'i, Moloka'i, Kaho'olawe, and Ni'ihau).²⁹ Strikingly, Diamond concludes his account of Hawai'i's unification by speculating that, had Hawai'i enjoyed "a few more millennia" of cultural isolation, Kamehameha's proto-state might have emerged as "a full-fledged empire" powerful enough to control the rest of Polynesia.³⁰

---

[22] Diamond, *Guns*, 276-78.
[23] Ibid., 54.
[24] Ibid., 64-5.
[25] Ibid., 64.
[26] Ibid., 64.
[27] Jim Dator, "New Beginnings Within a New Normal for the Four Futures," *Foresight* 16, no. 6 (2014): 498.
[28] Diamond, *Guns*, 64.
[29] Ibid., 64-65.
[30] Ibid., 66.

Although Mitchell follows Diamond's lead in using remote islands as the cipher through which to yoke an ecocidal past to a climate-changed future, "Sloosha's Crossin'" calls attention to the genocidal consequences of colonial violence in the "formation and maintenance" of Empire.[31] To underscore the impact of colonial opportunism absented from Diamond's account, "Sloosha's Crossin'" commences with a defamiliarized portrait of Captain Cook's introduction to and mistreatment of locals on Hawai'i's Big Island (1779), then proceeds with the return of Kona tropes to Waipi'o after the "Battle of Koapapa'a" (1790) and "Battle of Kepuwaha'ula'ula" (1791). "Sloosha's Crossin'" then concludes by dramatizing the pivotal alliance of Captain Vancouver with Chief Kamehameha (1793), in which the latter ceded Big Island to Great Britain in exchange for advanced weaponry, thus sealing Kamehameha's success in seizing Hawai'i's Big Island and, later on, Maui (1794). In the harrowing vignette that introduces "Sloosha's Crossin,'" Kona warriors kill Zachry's father near the Hiilawe Falls of the Waipi'o river.[32] Zachry's traumatic encounter corresponds with the regrouping of Kamehameha's consort following the "Battle of Koapapa'a." After attempting to seize Maui in the "Battle of Ka'uwa'iau," Kamehameha returned to Big Island's Kawaihae district (1790), where he was ambushed by his cousin in the Pa'auahau forest northeast of Honoka'a. Here, Kamehameha expelled his cousin with canons previously deployed in Maui.[33] After retreating east to Kūka'iau in the "Battle of Koapapa'a," Kamehameha returned to the Kohala district through Waipi'o, while his cousin retreated south to Ka'ū.[34] While recounting his harrowing encounter with Kona raiders, Zachry notes that his childhood recalls that of his father, who similarly outwitted Kona slavers in the district of Mo'okini as a boy.[35] This fleeting intergenerational reference corresponds with the actions of Kamehameha's uncle, who relocated the royal court to Mo'okini (1750) – a remote district on Hawai'i's Big Island purported to be the birthplace of Kamehameha.[36]

---

[31] Michael Wilcox, "Marketing Conquest and the Vanishing Indian," *Journal of Social Archaeology* 10, no. 1 (January, 2010): 98.
[32] Mitchell, *Cloud,* 251.
[33] William R. Castle, *Hawaii Past and Present* (New York: Mead Dodd, 1917), 34.
[34] Samuel M. Kamakau, *Ruling Chiefs of Hawaii* (Honolulu, Hawaii: Kamehameha Schools Press, 1992), 350.
[35] Mitchell, *Cloud,* 253.
[36] Kamahameha's birth was initially thought to have coincided with the transit of Halley's comet (1758) to which Mitchell alludes through the recurrence of a "comet-shaped" birthmark (*Cloud,* 85; 124; 373; 319).

Just as Mitchell's speculative skirmishes correspond to Kamehameha's unification of Hawai'i's Big Island (1793), Mitchell's imagined incursion of Prescient traders are modelled in composite after the introduction of eighteenth-century American and European seafarers to historical Hawai'i. In his account of thirtieth-century Ha-Why, Mitchell reimagines white American and European carriers of Eurasian diseases as "brewy-brown" Prescients immune to "redscab sickness."[37] Archival evidence suggests that Cook's third expedition knowingly introduced syphilis, gonorrhea, tuberculosis, and influenza to Hawai'i's Big Island with the long-term effect of decimating its local population.[38] Moreover, Mitchell's middle-aged Meronym offers an inverted portrait of Captain James Cook, who first spied the "Sandwich Islands" (1778) at fifty years of age. Just as Duophysite charges Meronym with surveying Ha-Why's Big I for the purpose of land and resource capture,[39] the British Admiralty tasked Cook with reporting sites suitable for British colonization while charting the Northwest Passage in 1776.[40] The manner in which Mitchell's seafarers anchor at "Flotilla Bay" to trade ironware for provisions[41] recalls Cook's final expedition, which anchored in Kealakelua Bay twice in quick succession (1779): first for one month to gather provisions and record astronomical observations; then for less than a week to repair the broken foremast of the *Resolution*.[42] When asked to identify the location of her "home valley," Meronym replies that she hails from an island absent from seafarers' maps – the uncharted islet of Prescient I found "far-far in the northly blue," north of "Ank'ridge" and "Far Couver."[43] Meronym's description adheres to the route of Cook's third expedition, in which Hawai'i supplied the necessary provisions for Cook to traverse the "northly blue" past the Juan de Fuca Strait, through Nootka Sound, north of Vancouver, and across the Bering Strait to Cook's Inlet, north of Anchorage.[44] In contrast to Zachry, who comes to worship Meronym as a "deity" after a brief period of resentment and distrust, eighteenth-century Hawaiians grew increasingly disenchanted with Cook

---

[37] Ibid., 259-260.
[38] Diamond, *Guns,* 214.
[39] Mitchell, *Cloud,* 310.
[40] Vanessa Collingridge. *Captain Cook* (London, United Kingdom: Ebury Press, 2003), 327.
[41] Mitchell, *Cloud,* 259-261.
[42] Gavan Daws, *Shoal of Time* (New York: Macmillan, 1968), 8; 16.
[43] Mitchell, *Cloud,* 263; 311-312.
[44] Derek Hayes, *Historical Atlas of the Pacific Northwest* (Seattle, Washington: Sasquatch Books, 1999), 42-43.

after his crew repeatedly abused their hospitality.[45] Strikingly, Meronym's steadfast refusal to supply the Valleysmen with "smart gear" or "spesh weapons" (259-60) chimes with Cook's general ethos of non-interference – an ethos later abandoned by an opportunistic Captain George Vancouver.

In the vignette that sees Zachry guide Meronym to the summit of Mauna Kea to explore its "observ'trees," Zachry briefly falls under the influence of Old Georgie, who unsuccessfully tries to tempt Zachry to doubt Meronym's good intentions anew.[46] Here, Old Georgie functions not only as a supernatural embodiment of evil and death but also as a semi-veiled allusion to Captain George Vancouver, who supplied Kamehameha with "European guns and ships" sufficient to conquer the largest islands in the archipelago.[47] Moreover, Zachry's victory over Old Georgie outside the "Old'un temple" corresponds with Kamehameha's construction of the Puʻukoloā heiau temple (1791) in the district of Kawaihae, northwest of Waimea.[48] In a ranch house nestled in the foothills of Puʻukoloā heiau, Vancouver gifted Kamehameha with cattle (1793) in exchange for permission to construct an astronomical observatory on the site.[49] The following winter (1794), Vancouver's midshipman became the first European to ascend to *Mokuaweoweo* (Mauna Loa's summit) while Vancouver negotiated Big Island's accession to Great Britain. In exchange for ceding Kamehameha's holdings to the Union Jack, Vancouver directed the construction of a thirty-foot warship stocked with artillery and a man-o-war fitted with cannons – all of which proved pivotal in Kamehameha's capture of Oʻahu (1795) in the "Battle of Nuʻuanu."[50] In the latter half of "Sloosha's Crossin,'" Meronym and Zachry are shown escaping to Ikat's Finger after Mitchell's Kona raid a central trading post in Honokaʻa.[51] While passing through his home valley, Zachry slits the throat of a napping looter – a gruesome episode that recalls the ravaging of villages in Waipiʻo following the arrival of Kamehameha's offshore rivals.[52] Once competing chiefs learned of Kamehameha's ambition to unify not only Big Island but the entire archipelago,

---

[45] Marshall Sahlins, "Captain Cook at Hawaii," *Journal of Polynesian Society* 98, no. 4 (December, 1989): 384.
[46] Mitchell, *Cloud*, 294-295.
[47] Diamond, *Guns,* 64.
[48] Van James, *Ancient Sites of Hawaiʻi* (Honolulu, Hawaii: Mutual Publishing, 1995), 133.
[49] Manley Hopkins, *Hawaii* (New York: D. Appleton and Co., 1869), 83.
[50] Speakerman Cummins and Rhoda Hackler, "Vancouver in Hawaii," *Hawaiian Journal of History* 23 (1989): 61.
[51] Mitchell, *Cloud,* 303; 309-11.
[52] Ibid., 316.

warriors from Oʻahu and Kauaʻi invaded Waipiʻo (1791), where they raided, looted, and destroyed villages near Zachry's home valley.[53] In the conclusion to "Sloosha's Crossin,'" Zachry and Meronym are last seen aboard Prescient kayaks bound for Maui.[54] This final vignette obliquely refers to the decisive victory that brought Big Island under Kamehameha's control. Near the basin of the Waipiʻo Valley, off the coast of Pali Hulaʻana, Kamehameha staged the epic sea battle of "Kepuwahaʻulaʻula."[55] At this point, Kamehameha's army – totaling sixteen thousand warriors and twelve hundred war canoes – constituted the largest in Hawaiian history.[56] Kamehameha swiftly seized Maui (1794), Lanaʻi, Molokaʻi, and Oʻahu (1795), then quashed an uprising on Big Island (1796), before gaining control of Kauaʻi and Niʻihau (1810), thereby unifying the Hawaiian archipelago into a single kingdom, which Kamehameha controlled from Oʻahu until his death (1819).[57]

Although Mitchell's speculative enactment of Hawaiʻi's unification appears to dramatize history's "default" trajectory as one in which peaceable communities are riven by acts of "genocide and enslavement" to sustain the demands of Empire, Mitchell's speculative portrait of Ha-Why disputes Diamond's characterization of the archipelago as a "pristine proto-State" by emphasizing the intrusive presence of British seafarers who, in supplying advanced weaponry to enterprising chiefs, enabled the expulsion of Hawaiʻi's windward communities.[58] Mitchell circuitously suggests that Diamond's analysis overlooks how complexity, flux, and historical remediation may enable or curb human predation, as evidenced when Mitchell peoples his novel's 1930s Belgium with a satiric avatar for Jared Diamond. In "Letters for Zedelghem," Mitchell introduces the reader to Morty Dhondt: a "multi-lingual" diamond merchant stationed in "Bruges and Antwerp" who delights in kicking "metaphysical football[s]" with Frobischer's calculating employer.[59] Though Dhondt makes but a brief appearance in *Cloud Atlas*, his address to Frobischer upon their return from Zonnebeke cemetery encapsulates the basic premises that underpin *Guns,*

---

[53] S.L. Desha, *Kamehameha and his Warrior Kekūhaupiʻo* (Honolulu: Kamehameha Schools Press, 2000), 296.
[54] Mitchell, *Cloud*, 324.
[55] Desha, *Kamehameha*, 299.
[56] Castle, *Hawaii*, 34.
[57] Diamond, *Guns*, 64.
[58] Berthold Schoene, "The World Begins Its Turn with You," in *The Cosmopolitan Novel*, ed. by Berthold Schoene (Edinburgh: Edinburgh University Press, 2009), 117; Diamond, *Guns*, 282.
[59] Mitchell, *Cloud*, 61.

Germs, and Steel. Here Dhondt maintains that the will to power is "the backbone to human nature;" that the "nation-state is merely human nature inflated to monstrous proportion;" that resource capture and war are "eternal companions" to humanity; that diplomacy merely "mops up war's spillages [and] legitimizes its outcomes;" and that, given the arsenal available to affect destruction, *Homo sapiens* will be "snuffed out" before the end of the century.[60] Although Dhondt's analysis grossly lacks Diamond's nuance, Dhondt's observations draw from claims Diamond posits in his comparative analysis of Rēkohu and Hawai'i. Here Diamond insists that human predation necessitates an ethos that resists ideological heterodoxy: that neither the ethical benefits of egalitarianism nor the economic sustainability of enviro-materialist stewardship are sufficient to "speak" to the logic of Empire. Instead, Diamond proposes that human predation is curbed by environmental limitation or authoritarian decree.[61] Furthermore, Diamond suggests that environments conducive to egalitarianism remain free of predation only through cultural isolation enabled by geographical obscurity.[62] While geophysical and cultural encapsulation may temporarily inoculate communities against predation, Diamond insinuates that, once breeched, circumscribed environments provide the ideal medium for extreme predation to flourish.[63] To the extent that organized religion plays a role in perpetuating human predation, Diamond contends that religious orthodoxies emerge only after state formation to buttress the centralization of authority. Citing Hawai'i's example, Diamond posits that organized religions function to justify the "transfer of wealth" from commoners to élites claiming "divine descent."[64] The mutually beneficial alliance of ruler and priest is particularly apparent, Diamond contends, in eighteenth-century Hawaiian chiefdoms, which saw *kahuna* (local priests) collect tribute from each district to support the construction of ever-larger *heiau* (temples) purposed to sanctify the authority of the *ali'i* (hereditary chiefs).[65]

Mitchell's speculative account of Hawai'i's recent past implicitly rejects Diamond's two-fold suggestion that human predation is the inevitable consequence of "disease-dust and fire-arms" and that organized religion is but a by-product of statehood.[66] Although Mitchell's Kona both outnumber and outperform Big I's Valleysmen, Mitchell attributes the

---

[60] Ibid., 462.
[61] Diamond, *Guns*, 55.
[62] Ibid., 56.
[63] Ibid., 59.
[64] Ibid., 278.
[65] Ibid., 278.
[66] Mitchell, *Cloud,* 509; Diamond, *Guns*, 278.

windwards' defeat to a perspectival deficit. Ha-Why's windward communities are prone to enslavement, Mitchell posits, because the Valleymen mistakenly presume the Kona to be indomitable, as evidenced when Zachry comments that the "Kona'd' o'done to the Valleys what happened sooner or later anyhow."[67] Conversely, Mitchell suggests that the Kona maintain their suzerainship by taking for granted that the windwards are "freakbirths" predestined to serve the Kona as the "true inheritors" of Ha Why.[68] In this respect, Mitchell's Valleysmen and Kona are both complicit in creating communal identities predicated on fictions of tribal purity. In contrast, Ewing of "Pacific Journal" contends that "history admits no rules," only outcomes motivated by belief:

> Belief is both prize & battlefield, within the mind & in the mind's mirror, the world. If we believe humanity is a ladder of tribes, a colosseum of confrontation, exploitation & bestiality, such a humanity is surely brought into being [...] If we *believe* that humanity [can] share this world as peaceably as the orphans share their candlenut tree,[69] such a world will come to pass. I am not deceived. It is the hardest of worlds to make real.[70]

Mitchell goes on to explain that a universal theory of human predation cannot be "divvied" in terms of "tribes or b'liefs" or "mountain ranges" alone.[71] Tribal, colonial, and transnational manifestations of predation may be historically recursive; however, Mitchell's macronarrative insists that humans remain a protean species prone just as much to historical permutation as to reiteration. Indeed, Mitchell's novel insists that human collectives are not monolithic but "hydra-headed;" that "ev'ry [beautsome] heart" suggests "the savagery o'jackals;"[72] and, that every human is "Civ'lized and Savages both, yay."[73] In effect, Mitchell's masterpiece posits that human propensity is manifold, prone to flux, and characterized by

---

[67] Mitchell, *Cloud*, 296.
[68] Ibid., 291; 305.
[69] This reference recalls an earlier episode in which Ewing, from a hospital bed in Hawai'i, spies orphans scaling the branches of an "obliging" candlenut, each helping their playmate to ascend (527). In eighteenth-century Hawai'i, candlenuts were burnt to provide light and used in ceremonies as a symbol of wisdom and peace (Krauss 51).
[70] Ibid., 528.
[71] Ibid., 319.
[72] A subtle allusion to the British sloop the *Jackall*, which supplied arms to warring communities on O'ahu between 1792 and 1794 (Stokes, "Honolulu," n.p.).
[73] Mitchell, *Cloud*, 319.

internal contradiction – as Zachry confirms in his comment that an individual will "b'lief in a mil'yun diff'rent b'liefin's" at once if "jus' one" may be of aid.[74] In the vignette that concludes Zachry's tale, the injured goatherd looks at the sky from Duophysite's kayak to ponder whether "souls cross the skies of time" like clouds cross "the skies o' the world."[75] Zachry concludes that, just as the hue or shape of a cloud may shift over time but remain a cloud in kind, so too the contents of the human soul.[76] It is belief alone, Zachry reasons, that effects such shifts in the soul, for belief constitutes the "east'n'west," the "map," and even those uncharted regions "b'yonder" the map's edge.[77] In taking stock of the expansive dimensions of the human soul, Zachry offers this devastating blow to Diamond: "Who can say where the cloud's blowed from or who the soul'll be 'morrow," for belief is both the soul's "compass" and its "atlas."[78] Ultimately, *Cloud Atlas*' vast array of recursive realms demonstrates that the soul avails itself to human perception only as a set of transhistorical and transnational "variations" that resist efforts to map fixed or universal "essences."[79] Thus, when Zachry asks Meronym how the "true true" differs the "seemin' true," Meronym replies that, when bestirred from the "swamp of dissent," the "true true is presher'n'rarer'n *diamonds.*"[80] In refuting Diamond's premise that redistributive economies born of food surpluses necessitate the stratification of society islands into complex kleptocracies sustained through ceremonial works and wars of conquest, Mitchell's novel demonstrates that the only truth to be gleaned from the study of human collectives is that the totalities on which human predation relies (total purity, total difference, and total circumscription) are but illusions sustained through recursive acts of reception and transmission. By extension, Mitchell's novel insists that human collectives may act to disrupt the genre of thought that underwrites enviro-materialist theories of human predation.

Conclusion

Ultimately, in transposing eighteenth-century Hawai'i onto a speculative thirtieth-century Ha-Why, Mitchell's *Cloud Atlas* gestures toward a constellation of contemporary climate fictions that similarly

---

[74] Ibid., 279.
[75] Ibid., 318.
[76] Ibid., 324.
[77] Ibid., 318.
[78] Ibid., 324.
[79] O'Donnell, "Introduction," 70.
[80] Mitchell, *Cloud*, 287; 17; 288 (italics mine).

submerge and compress historical referents to discredit the determinacy of being-in-time. Indeed, Mitchell's partially veiled allusions have little to do with the author's alleged reticence to speak on behalf of colonized communities.[81] Rather, Mitchell's referents advance the position that the past is necessarily remediated according to cultural norms that install the illusion of a future in which a "purely predatory world [shall consume] itself."[82] Mitchell suggests, both within "Sloosha's Crossin'" and elsewhere, that human predation is not merely the predictable upshot of food surpluses and geopolitical circumscription. Rather, predation is a genre of thought that insists upon its own intractable inevitability. The current conventions that encapsulated communities use to navigate being-in-time may be transcended, Mitchell posits, only once humans "can conceive of doing so" – for just as condemned buildings are easily reduced to "rats' nests & rubble," so too may outdated conventions of thought.[83] On several occasions, Mitchell publicly expressed his concern with unchecked climate change and the advent of peak oil. Addressing an auditorium of bilingual teenagers enrolled in Madrid's British Council School, Mitchell compared global reliance on petroleum to a lethal addiction. Elsewhere, Mitchell disclosed that the prospect of petroleum depletion frightens him both "as a writer and as a dad."[84] While Mitchell remains optimistic that humans might elect leaders with "integrity," he also acknowledges that human civilizations might end as all things end: "Some [civilizations] collapse in slow motion. Others collapse almost overnight."[85] Although Mitchell does not consider himself an "explicitly political" author, he recognizes that any attempt to "write about the world with integrity" necessitates that politics enter prose; whereas some authors invite politics in "by the front door," Mitchell prefers to let politics in through "the back."[86] It is perhaps for this reason that *Cloud Atlas* makes no direct reference to fossil fuel dependency, nor to the cataclysmic threat posed by climate change, though both clearly weigh on Mitchell's mind. Notably, *Cloud Atlas* takes its namesake from the pictograms meteorologists use to predict the weather.[87] Mitchell's Ha-Why

---

[81] Martin Paul Eve, "The Conservatism of *Cloud Atlas*," n.p.
[82] Mitchell, *Cloud,* 528.
[83] Ibid., 479; 506.
[84] William Stephenson, "'Moonlight Bright as a UFO Abduction'" in *David Mitchell: Critical Essays*, ed. Sarah Dillon (Canterbury, United Kingdom: Gylphi, 2011), 244.
[85] David Mitchell, as cited in James Kidd, "Time and Again," *National* (October 9, 2014): n.p.
[86] Morère, "*Cloud*," 287.
[87] O'Donnell, "Introduction," 79.

indirectly attributes civilizational collapse to territorial wars sparked by mass displacements resulting from rising seas.[88] More precisely, "Sloosha's Crossin'" commences with floodwaters submerging Honoka'a market[89] and the ill-timed arrival of plague-ridden Prescient refugees,[90] whereas the narratological climax of "Sloosha's Crossin'" occurs in the ancient "observ'tree" presently used to monitor global atmospheric CO_2 levels.[91] At least obliquely, *Cloud Atlas* suggests that, just as the windward communities of Ha-Why had nowhere to flee once conditions on Big Island deteriorated, twenty-first-century humans may soon find themselves marooned on hostile terrain once climate change renders the Earth an equally inhospitable site. Ultimately, in its expansive portrait of human predation, Mitchell's novel attributes Anthropogenic climate-change to an imaginative deficit: the seeming inability of humans to conceive of a genre of being-in-time exuberant enough to overtake a recursive fiction of Empire. Although Mitchell suffers no illusion that generating a genre of thought sufficient to overcome ecocidal predation will require nothing less than realizing the "hardest possible world to make real," Mitchell grants that the non-inevitability of any particular future furnishes humanity with a "flea o' hope."[92]

References

Abell, Stephen. "How to Get the Ker-Bam." Review of *Cloud Atlas*, by David Mitchell. *Times Literary Supplement* 5265 (February 27, 2004): 21-22.

Begley, Adam. "The Art of Fiction No. 204." *Paris Review* 193 (Summer 2010) https://www.theparisreview.org/interviews/6034/the-art-of-fiction-no-204-david-mitchell

Bracke, Astrid. "The Contemporary English Novel and its Challenges to Ecocriticism". In *Oxford Handbook of Ecocriticism*, edited by Greg Garrard, 428-431. (Oxford: Oxford University Press, 2014).

Brown, William J. *Syphilis and Other Venereal Diseases*. Cambridge, Massachusetts: Harvard University Press, 1970.

Castle, William R. *Hawaii Past and Present*. New York: Mead Dodd, 1917.

---

[88] Mitchell, *Cloud*, 249-325.
[89] Ibid., 249.
[90] Ibid., 310.
[91] Charles D. Keeling et al. "Atmospheric Carbon Dioxide Variations at Mauna Loa Observatory, Hawaii," *Tellus* 28, no. 6 (January 1, 1976): 538.
[92] Mitchell, *Cloud*, 528; 319.

Collingridge, Vanessa. *Captain Cook: The Life, Death, and Legacy of History's Greatest Explorer*. London, United Kingdom: Ebury Press, 2003.

Correia, David. "F**k Jared Diamond." *Capitalism, Nature, Socialism* 24, no.4 (2013): 1-6. doi: 10.1080/10455752.2013.846490.

Cummins, Speakman, and Rhoda Hackler. "Vancouver in Hawaii." *Hawaiian Journal of History* 23 (1989): 31-65. http://hdl.handle.net/10524/121.

Dator, Jim. "New Beginnings Within a New Normal for the Four Futures." *Foresight* 16, no.6 (2014): 496-511. http://dx.doi.org/10.1108/FS-09-2013-0046.

Daws, Gavan. *Shoal of Time: A History of the Hawaiian Islands*. New York: Macmillan, 1968.

Desha, S.L. *Kamehameha and his Warrior Kekūhaupi'o*. Honolulu, Hawaii: Kamehameha Schools Press, 2000.

Diamond, Jared. *Guns, Germs, and Steel: The Fates of Human Societies*. New York: W.W. Norton & Co., 1997.

Eve, Martin Paul. "The Conservatism of *Cloud Atlas*." Accessed 21 June, 2015. eve.gd/2015/06/21/the-conservatism-of-cloud-atlas/.

Ferguson, Paul. "'Me Eatee Him Up': Cannibal Appetites in *Cloud Atlas* and *Robinson Crusoe*." *Green Letters* 19, no.2 (2015): 144-156. doi: 10.1080/14688417.2015.1022869.

Hardt, Michael and Antonio Negri. *Empire*. Cambridge, Massachusetts: Harvard University Press, 2001.

Hayes, Derek. *Historical Atlas of the Pacific Northwest: Maps of Exploration and Discovery*. Seattle, Washington: Sasquatch Books, 1999.

Hopf, Courtney. "The Stories We Tell: Discursive Identities Through Narrative Form in *Cloud Atlas*". In *David Mitchell: Critical Essays*, edited by Sarah Dillon, 104-126. (Canterbury, United Kingdom: Gylphi, 2011).

Hopkins, Manley. *Hawaii: The Past, Present, and the Future of its Island-Kingdom*. New York: D. Appleton and Co., 1869.

James, Van. *Ancient Sites of Hawai'i*. Honolulu, Hawaii: Mutual Publishing, 1995.

Juvik, Sonia P. *Atlas of Hawai'i*. Honolulu, Hawaii: University of Hawai'i Press, 1998.

Kamakau, Samuel M. *Ruling Chiefs of Hawaii*. Honolulu, Hawaii: Kamehameha Schools Press, 1992.

Keeling, Charles D. et al. "Atmospheric Carbon Dioxide Variations at Mauna Loa Observatory, Hawaii." *Tellus* 28, no.6 (January 1, 1976): 538-551. doi: 10.3402/tellusa.v28i6.11322.

Kidd, James. "Time and Again: Interview with David Mitchell." *National* (October 9, 2014). www.thenational.ae/arts-culture/time-and-again-the-critically-acclaimed-novelist-david-mitchell-on-life-death-and-everything-in-between-1.589780?videoId=5771275459001.

Krauss, Beatrice H. *Plants in Hawaiian Culture*. Honolulu, Hawaii: University of Hawai'i Press, 1993.

Leith, Sam. "A Literary Houdini." *Telegraph* (February 24, 2004) www.telegraph.co.uk/culture/books/3612689/A-literary-houdini.html.

Lindner, Oliver. "Postmodernism and Dystopia: David Mitchell, *Cloud Atlas*". In *Dystopia, Science Fiction, Post-Apocalypse*, edited by Eckart Voights and Alessandra Boller, 363-377. (Trier: Wissenschaftlicher: Verlag Trier, 2015).

Mason, Wyatt. "David Mitchell the Experimentalist." *New York Times Magazine* (June 25, 2010).
https://www.nytimes.com/2010/06/27/magazine/27mitchell-t.html.

Mitchell, David. *Cloud Atlas*. London: Sceptre, 2004.

—. "On Historical Fiction." *The Thousand Autumns of Jacob de Zoet*. London: Sceptre, 2011.

Morère, Julie. "*Cloud Atlas*: A Novel Questioning Mitchell's Avowed Lack of Commitment to the Real World." *PU de la Méditerranée* (2010): 285-296.

O'Donnell, Patrick. "Introduction: Many Worlds, Real Time". In *A Temporary Future: The Fiction of David Mitchell*, edited by Patrick O'Donnell, 1-21. (London, United Kingdom: Bloomsbury, 2015).

Parker, Jo Alyson. "David Mitchell's *Cloud Atlas* of Narrative Constraints and Environmental Limits". In *Time: Limits and Constraints*, edited by Jo Alyson Parker, Paul Harris, and Christian Steineck, 201-218. (Boston: Brill, 2010). http://doi.org/10.1163/ej.9789004185753.i-378.79.

Sahlins, Marshall. "Captain Cook at Hawaii." *Journal of Polynesian Society* 98, no.4 (December, 1989): 371-423.
http://www.jps.auckland.ac.nz/document/?wid=3478.

Schoene, Berthold. "The World Begins Its Turn with You, or How David Mitchell's Novels Think". In *The Cosmopolitan Novel*, edited by Berthold Schoene, 97-126 (Edinburgh: Edinburgh University Press, 2009).

Sorlin, Sandrine. "A Linguistic Approach to David Mitchell's Science-Fiction Stories in *Cloud Atlas*." *Miscelánea* 37 (2008): 75-89.
https://halshs.archives-ouvertes.fr/halshs-01271546/document.

Stephenson, William. "'Moonlight Bright as a UFO Abduction': Science Fiction, Present-Future Alienation and Cognitive Mapping" In *David*

Mitchell: Critical Essays, edited by Sarah Dillon, 225-246 (Canterbury, United Kingdom: Gylphi, 2011).

Stokes, John Francis Gray. "Honolulu and Some New Speculative Phases of Hawaiian History." *Annual Report* 42 (Honolulu, Hawaii: Hawaiian Historical Society, 1933).

Trexler, Adam. *Anthropocene Fictions: The Novel in a Time of Climate Change*. Charlottesville, Virginia: University of Virginia Press, 2015.

Wilcox, Michael. "Marketing Conquest and the Vanishing Indian: An Indigenous Response to Jared Diamond's Guns, Germs, and Steel." *Journal of Social Archaeology* 10, no.1 (January, 2010): 92-116. doi: 10.1177/1469605309354399.

Wright, Ronald. *A Short History of Progress*. Toronto: Anansi Press, 2004.

Chapter Eight

The 'Dying' Earth in the Postnatural Age of the Anthropocene: An Ecological Reading of Amitav Ghosh's "Cli-Fi" Sequel

Risha Baruah

Introduction

The contemporary period of the Anthropocene has witnessed a motivated denaturalization of the environment through the logic of domination insulated by the nature-culture binary as channelized by ideals like anthropocentrism, egoism and humanism. While these social ideologies throughout history have manipulated our interactions and interpretations of nature with an aim to control, exploit and alter the natural world, it has also justified the endless invasive and aggressive interventions of humans that have 'unnaturally' hybridized the environmental as well as the geological imprints of the planet. This configuration further led to the significant crippling of the eco-centric relationship between humans and nature which thereafter exposed the ecosystem to greater risks, instabilities and vulnerabilities that resulted in a heightened increase of frequency and intensity of ecological catastrophes in the Anthropocene. This notion has been addressed by several prominent scholars through the new emergent literary form "cli-fi" with an aim to direct responses for a better understanding, embracing and shaping of the impending apocalypse. To this end, the chapter shall attempt to investigate the literary efforts of Amitav Ghosh through his "cli-fi" sequel comprising of *The Hungry Tide* (2004) and *Gun Island* (2019) that shall be considered as a singular fiction. Towards this consideration, Ghosh extensively engaged with approaches like ecocriticism and Anthropocene to understand the complex and dire social and environmental situations that have led to the aggressive destabilizing and declining health of the planet as witnessed in the postnatural world. While this ecological effort may expose a general sense of fear, anxiety, insecurities, displacement, and hostility as experienced by the contemporary generation, its central intention has been to reorient the ongoing ecophobic towards the ecological crisis from violence, denialism and repeated terror to *timely* warning, awareness, activism and stewardship for a productive management of the ecocide that is pushing the ecosystem to the brink of an apocalypse.

Climate Fiction and Amitav Ghosh

As noted through the Great Acceleration, the contemporary society has been exposed to innumerable environmental catastrophes that have been human engineered through the heightened sense of anthropocentrism, high modernism, international capitalism, unplanned development, urbanism and industrialism which manipulated the shaping of the 'second nature' as evident in the post-natural world of the Anthropocene. This tendency for Bookchin as cited by Manes gave humans "not only the *right* but the *duty* to alter, shape and control 'first nature' (the nonhuman world)" (1996, 23). Towards this end, the idea of 'second nature' appears to be tainted by a materialist appraisal of socionatural relations. In this frame of 'human-centric elitism' and egoism as endorsed by the western ideology of anthropocentrism, we also witness the mechanism of bio-power politics as managed by the nature-culture binary which has played a pivotal role in fracturing the equilibrium between humans and the environment. This motivated interaction of humans with the natural world has led to the presence of "*only* touched nature" for there is "no part of the earth left untouched by man" with the continual hybridization of the ecology in Anthropocene (emphasis added) (Maldonado, 2015, 34, Soper, 2000, 124 and Schwagerl, 2014, 110).

Such a tendency further increased the detachment of humans from nature which resulted in a systematic and endless series of aggressive and invasive interactions between them that has unfortunately led humankind to the contemporary geological epoch. This has furthered the fundamental tension in our society as the idea of 'nature' as fabricated by the cultural agencies has not only led to its limitation as a social construct that alienated humans from their physical environment but also led to its tremendous, irreparable and dangerous degeneration by humans who have ethically failed to recognize "nature as an *equal* partner" in their existence, survival and sustenance (emphasis added) (Krebber in citing Latour, 2011, 324). Subsequent to this, nature has conveniently been marginalized as the "absent [and silenced] *other*" (emphasis added) (Nimmo, 2011, 77). This mechanism of 'othering' through the social dictum of anthropocentrism has led not only to the domination of the external environment by humans but also led to "the domination of the inner nature of humans, which [further led] to the domination of humans over humans" (Krebber, 2011, 324 and Clark, 2013. 79). Adding to this, Rob Boddice claimed that anthropocentrism not only caused problems "for non-humans, but also for humanity" which made its intent a disservice to mankind for it conditioned and victimized humans as well as the nonhuman (2011, 17). Such a lopsided tendency led to an intentional and systematic legitimization of the ecology for the singular benefits of humans. To this end, nature was considered as a 'dangerous and foreboding' *presence* in the Anthropocene period which has often been encountered with repeated terror, anxiety, anger, hostility and denialism to its emerging ecocide. In fact, the rigid fixation of ecology in extreme polarity with culture in social discourses has led to a deepened ecophobia that has eventually resulted in an "ecological blindness" in humans who are driven by an irrational desire to 'master and control' the ecosystem (Hiltner, 2015, xv and Deyo, 2018, 195). Caught in this social fabrication, humans have often engaged in aggressive as well as invasive anthropocentric attitudes and activities that have unnaturally altered the functioning of the ecosystem. Keeping to this, the 1990s witnessed an increased ecological engagement of social and environmental crisis through climate change that soon "began to make its way into the cultural [and literary] imagination". Such a response by contemporary efforts has led to a widened understanding of the ongoing planetary transformation and how it has been affecting individuals as well as regions of the Earth (Heise, 2008, 205). For this purpose, ecocriticism and Anthropocene have attempted to understand the complex relationship and interaction between humans and the environment which has largely been anthropocentrically scaled (Glotfelty, 1996, xix, Purdy, 2015, 3, and Bates, 2018, 171).

To this end, both ecocriticism and Anthropocene have collaborated to understand the systematic tendencies that have 'tremendously and dangerously' altered the ecological and planetary imprints of the ecosystem as witnessed in the contemporary postindustrial and technocratic period of the Anthropocene. This lopsided equation has resulted in a visible tension between humans and the environment as humans have continually engaged to control, manage, and dominate nature and its resources throughout history. In this sense, humanity has become "a decisive geological and climatological force... through the natural becoming, as it were, dangerously out of bounds, in extreme and unprecedented weather events, ecosystems being simplified, die-back, or collapse". In this context, "the climate of our planet [appears to be] under our influence" as humans have become both the 'master' and "*end* of nature" (emphasis added) (Singer, 2011, 245). However, this trend in climate change dates back to Industrial Revolution that emphasized on carbon economy which led to the eventual reorientation of the environment with a 'mechanical view' that attempted to justify the endless exploitation of the ecology and its natural resources which were commodified and reduced to "raw materials" as directed by instincts of high modernism, cosmopolitanism, consumerism and contemporary globalised capitalism. Keeping this in consideration, climate change could be seen as a "cultural diagnosis of the tension rising from the nature/culture dichotomy as proposed by anthropocentrism" (Clark, 2013, 79-80 and Scranton, 2015, 22-23). Such an exercise has led to application of scientific, environmental and geological references to humanities and literary efforts with an aim to include social conceptions codified by humans. In fact, it further allowed the cross-fertilization of traditional genres in literary, disciplinary and critical approaches with the increased engagement of ecocriticism and Anthropocene with climate change. Towards this end, 2007 witnessed several eminent ecocritical scholars like Timothy Morton, Ursula K. Heise, Scott Slovic and Richard Kerridge claiming climate change as the "central preoccupation of the field". In addition to this, Slovic not only critiqued "climate deniers" but also called for a boarder project to trace "climate science's role in society, investigating historical responses to climatic change and interpreting contemporary climate change discourse". Such an initiative directed the possibility of a merger between science, humanities and literature which subsequently led to the redefinition of nature as "a planetary interactive casual network operating across multiple scale levels, temporal and spatial" (Trexler, 2015, 18 and Adamson, 2013, 174-175). With the remapping of climate change in ecocriticism and Anthropocene, there has been a visible production of climate change fictions in the recent years which had prompted Dan Bloom to shorthand it to "cli-fi" which had subsequently "spurred the creation of a whole new genre of fiction" (Trexler, 2015, 8 and Sullivan, 2017, 28).

In addition, Garrard citing Thompson stated that the new literary genre of apocalypse has been closely associated to climate change and futuristic literature. The term has its origin in the Greek word 'Apocalyptein' which meant "to unveil". In this sense, the efforts of "cli-fi" appear as a "form of a revelation of the end of history" that has been largely filled with "violent and grotesque images" with its continual "struggle between good and evil". This made the literary form emerge "out of crisis" with a strategic aim to provide warnings, awareness, and activism (Garrard, 2007, 86 and Chakrabarty, 2015, 335). In this sense, the global phenomenon of climate change had furthered iterated the idea of apocalypse that was conceived through innumerable responses ranging from hopelessness, fear, anxiety and paranoia to denialism, violence and indifference which collectively failed to substantiate any productive attempt to mitigate the dire ecological problem. While this stance has been a visible challenge in the social world aided by scientific efforts, the literary task of representing climate crisis seems to be more evocative as it allowed a prolonged lingering on the imagination and sensibility of its readers to redraw their anthropocentrically scaled perceptions, choices and actions to an eco-centric structure mapped by tendencies like eco-cosmopolitanism, sustainability, stewardship, deep ecological commitments based on bio-centric ethics. To this end, apocalypse as managed and represented by ecocriticism as well as by Anthropocene studies aided in the formation of new "innovative and broadly conceptualized framework to rethink the relation between nature and culture, environment and society" which would thereafter refine human interactions with nature by "re-arranging the socionatural entanglement in a more enlightened, reflective way". Such a frame would not only "'liberate' nature, but it will protect the remaining natural forms in the context of a highly technological world that is rapidly in the making" (Maldonado, 2015, 85 and 124). In this sense, the genre of apocalypse in postcolonial and postmodern ecocriticism contained within itself the element of thriller, science and futurism that urged us to undertake ethical responsibility towards the ecosystem following the dismantling of the traditional nature/culture binary that was systematically endorsed by the western ideology of anthropocentrism that subtly aimed to justify the irreparable and irreversible damage caused to the planet.

Despite these efforts, nature has continued to be limited in its scope and representation in literary, social and theoretical efforts which reaffirms the idea of the 'death of nature' as advocated by McKibben Bill in his work, *The End of Nature* (1989) which traced the reductive capacity of nature with its intended instrumental and mechanical value. While such a tendency fuelled Man's egoism and exceptionality, it also widened the growing ambivalence and detachment between humans and nature which played a significant role in reorienting the previous reverential and eco-centric equation proposed by animism, paganism and totemism to the contemporary relationship of hatred, anger, vengeance, anxiety, fear and destruction as manipulated by the invasive interaction between humans and the ecology. This ecophobic response had reduced nature into a dangerous *presence* which resulted in the strategic management of the environment through hostility, denialism and a frantic ecocide. Owing to this drastic shift, there has been an extreme societal and ecological collapse as expressed through the apocalyptic rhetoric which according to Gabriel and Garrard had induced "debilitating apathy, rather than engagement and participation" (2012, 119). In this regard, human failed to extend moral consideration and ethical responsibility towards the environment which resulted in an unnatural acceleration of the environmental genocide with dangerous alteration of the 'naturescape' of the planet. Such a trend appeared to be an outcome of the hierarchical dominion of man as claimed by "The Great Chain of Being" which allowed humans to control, manage and transform the environment to cater to the endless needs of humans. Within this framework, human(ity) has often played a motivated role whose agency has often been considered as an indispensable tool to *improve* the natural resource management and mitigate harmful effects of human activities.

Keeping these assumptions in consideration, the chapter shall attempt to explore social, economic and political entanglements with environment following the "material turn" which resulted in a series of unnatural calamities that has been addressed by Amitav Ghosh in his "cli-fi" sequel comprising his novels, *The Hungry Tide* (2004) and *Gun Island* (2019) which shall be considered as a singular fictional narrative in this analysis. In fact, Ghosh in his environmental efforts actively engaged with the approaches of ecocriticism and Anthropocene to understand the complex social and environmental situations that have led to the aggressive destabilizing and declining health of the planet. To this end, the western ideology of anthropocentrism shall be investigated as envisaged in the postnatural world. This engagement by Ghosh appeared as a crucial effort to provide a better understanding, embracing and directing of social responses to the impending apocalypse. For this purpose, he traced the tension arising from the divorce of humans from their natural environment as framed by the nature/culture binary. Such a configuration had not only polarized nature and culture in opposition but also justified the series of unprecedented calamities that had largely been human engineered. In fact, this trend grew out of the increased emphasis on human exceptionality, egoism and supremacy that led to the aggressive invasion of the environment by humans which had irreversible and dangerous outcomes that threatened the stability and security of the ecosystem. Following this, the unpredictable and threatened reaction of nature to anthropogenic actions had been 'unfairly' reduced into negative responses for there has been an unprecedented increase in the frequency and intensity of natural calamities like floods, cyclones, tsunamis, forest fires and storms that have not only destroyed life and property but also livelihood which subsequently posed several social and economic insecurities and risks. Along with these social, environmental and planetary concerns, Ghosh in both his "cli-fi" novels also attempted to trace the weakening of physical, economical, emotional and spiritual capacities of the indigenous communities in the Sundarbans. Towards this end, Ghosh considered factors like overpopulation, tourism, development, capitalism, urbanism and modernity as instigators that manipulated the cultural shift from an ecological view of nature to a materialistic and mechanistic view of the environment. Such a consideration led to an increased detachment and negligence of the regional ecology as well as the remote indigenous communities that are often cornered to the intended social periphery of power politics and social identities.

In this frame, Anthropocene through climate change had attempted to address several global phenomena that threaten to obscure local and regional variations in the 'socionatural' relations of the planet. To this, the local tribes of the region like the "ecological refugees" often resorted to resistance and violence in their interaction with nature (Ghosh, 2017, 4). This was because of the increased social, emotional and economical insecurities, risks and deaths that changed the traditional filial, reverential and conservational interaction between humans and the environment. While addressing this concern, Ghosh emphasized that the younger generation from the "diverse" archipelago were becoming *more* defiant, aggressive, and negligent to the ecological concerns and considerations. Towards this end, the indigenous tribes engaged in material access, control, management and (ab)use of the natural resources through activities like overfishing, hunting, mining, deforestation, tourism, urbanism, industrialism, modernism, capitalism, migration, pollution, smuggling, overpopulation, global warming, cross-breeding, mutilation of animals, increased use of toxic chemicals and unplanned development which aided the scope of a megafauna extinction of the planet. In fact, through these activities humans have not only altered the functioning of the ecosystem thereby making it more unpredictable and prone to frequent and intense calamities but also aimed to control and domesticate the 'naturescape' of Sundarban as seen through the capitalistic aspirations of Daniel Hamilton and the illegal immigrants of the island Morichjhapi who "*always* lived-by fishing, by clearing land and planting the soil" (emphasis added) (Ghosh, 2013, 262). This exhausted not only the resources but also the regenerative capacities of nature which subsequently had an adverse effect on the natural functioning of the planet.

This socio-ecological configuration was addressed by Ghosh in both *The Hungry Tide* and *Gun Island* to generate *timely* warning and awareness among readers that frequent restructuring of nature would inevitably result in the destabilisation of the equilibrium in the ecosystem. To this end, both Ghosh and McKibben considered natural calamities as occurrences that "are not without a man-made cause" for our responses, interactions as well as "our ideas about and representations of nature are *dangerously* anthropocentric" which has detoured most of our eco-centric efforts into superficial environmentalism. This unconscious yet motivated shift has caused invisible but irreparable harm and degeneration of the environment (emphasis added) (McKibben, 2003, xx, Hamilton and Jones, 2013, 46). To this end, Ghosh traced the collaborative efforts of anthropocentrically scaled factors that have changed the natural pattern of each of the islands in Sundarban which are "constantly being swallowed up by the sea; they're disappearing before our eyes" (2019, 18). This was mostly because climate change had consequently led to rise of global temperature as well as the rise of sea levels following the continual melting of glaciers. In fact, climate change has accelerated the frequency of catastrophes which has exposed the ecosystem to endless and dangerous vulnerabilities, stress and exhaustion thereby pushing the planet to an unprecedented ecological crisis. To this, McKibben added that climate change and pollution has made "every spot on earth man-made and artificial" as commonly witnessed in the contemporary period of the on-going 'postnatural' world (2003, 7). In acknowledgement of this dire crisis, Ghosh represented the Sundarbans as a powerhouse of myriad explored and unexplored ecosystems that have not only been exposed to degeneration but also its eventual endangerment with increased human invasions through unplanned development and uncontrolled migration triggered by urbanism, modernism, capitalism and industrialism that has disrupted the uniformity of patterns in the natural cycles and functioning. These social ideals for Ghosh reiterated the ambitious obsession of humanity for material aspirations as a suicidal act; for without a planet, there can be no future and existence of humankind. While, the alarming activity of climate change entangled cultural, urban, industrial and literary mediations to trace the emotional and ethical reconfigurations among humans for and against nature, it also engaged in global and regional reporting of irreparable degeneration and toxic contamination of the planet. Towards this end, Ghosh highlighted the alteration of the Sundarbans by Daniel Hamilton, the illegal migrants in the island of Morichjhapi, the government as well as the emerging industries and refineries as an integral cause of frequent calamities like storms, cyclones, forest fires, tsunamis and flooding which made the Sundarbans reach "a volume where its very variety had undergone a [devouring] change" thereby making the landscape vulnerable and threatened (Ghosh, 2013, 378-379). In fact, these ecological changes have often been responded to by heightened anger, fear and anxiety by the local communities towards nature and the nonhumans as their survival and sustenance was threatened.

Such an anthropocentrically scaled attitude has led to an unaccounted toxic contamination of the environment through agricultural, naval and industrial runoffs that have resulted in mutation of the behavioural and breeding pattern of nature and the nonhumans along with increased dead zones in the marine ecosystem. This problem was emphasised by Ghosh in his novel, *Gun Island* that addressed the concern over industrial run-offs which had resulted in the dead zones "growing at a phenomenal pace" as a chain reaction to the increase of toxicity in the marine ecosystem (2019, 95). While the condition appears to be essentially an oceanic concern, it has reached river bodies as there is no definite border separating them. This unnatural phenomenon has subsequently witnessed large scale stranded deaths of marine creatures. In addressing the concern through the cetologist Piyali Roy, Ghosh emphasised that such a condition was no longer a regional mishap but a global situation. In fact, the increased episodes of beaching of fishes, crabs and dolphins in frequent intervals have not only been a result of industrial waste but also has been an outcome of "man-made sounds- from submarines and sonar equipment" used for tourism, navigation and fishing purposes by the locals in the Sundarbans. Such activities further disoriented the navigation pattern of marine creatures like dolphins as mobilisation through 'echo' gets disrupted. In addition, the increased use of nylon nets, motor boats and steamers in the region has killed innumerable marine creatures 'unnecessarily and unintended' (2013, 99). This alarming condition of increased toxicity and aggressive invasion of the wildlife has not only adversely paralysed nature but also humans.

In addition, Ghosh in his "cli-fi" sequel addressed the social and ecological implications of deforestation. To this end, he investigated the intensive formation of 'urbanature' in the region of Sundarban where forested jungles have been increasingly encroached to give way to crowded townships that were filled with strangers who lived a simplistic life filled with hardships and social evils like poverty, human trafficking, prostitution, crimes along with emotional dejection, fear, anxiety, deprivation of fundamental rights and dignity. Towards this consideration, Ghosh also attempted to highlight the difference between rural and urban landscape as experienced by Kanai Dutt (in *The Hungry Tide*) and Dinanath Datta (in the *Gun Island*) as they both contemplated on the beauty, diversity and power of nature and eco-centrism in their limited but influential exposure of the Sundarbans to that of their sophisticated but artificial encounters with cosmopolitan cities like New Delhi and Venice respectively which were largely modeled on global capitalism and high modernism that fuelled the exploitation of natural resources for anthropocentric benefits. In fact, the increased complexity of modern lifestyle through activities like overconsumption, illegal trading, vivisection, overpopulation and pollution have *injudiciously* curtailed the healthy regeneration of the 'dying' ecosystem that has been anthropocentrically scaled with the increased capitalization of the environment. Such a motivated interaction and interpretation of the environment has further delayed and detained our deep commitments to ecological benefits, ethics and activism. This is because nature has often been considered as an *economy* that needs to be anthropocentrically measured, managed, controlled and (ab)used. In consequence to this, ecology has been dangerously endangered with irreparable imbalance of the ecosystem through intensive hybridization of the geological imprint of the planet. Such an unnatural alteration of the ecosystem has accelerated the natural evolution of the planet by making it more vulnerable and unstable. Following this detached temperament, humans' perceptions and knowledge of the natural world has largely been tainted by material and ambitious aspirations. In consequence of this illusionary and artificial trend, humans have modeled their biased notion of nature with superficiality and reductionism which has gradually blinded us to the seriousness of the ongoing environmental crisis that has been growing at exponential capacities. This situation, for Ghosh resulted not only in the social and ecological dislocation but also in the degeneration of nature and humans as seen through his "spiritual death" that further made him increasingly intolerant and "alienated, empty, without purpose and direction" in his social discourse as well as in his complex relationship with nature. Such a cultural replacement not only made humans "unaware of a connection with Nature" but also made nature "virtually *nonexistent* to his perceptions" (emphasis added) (Fromm, 1996, 33). This complex bio-political dimension in humans' interaction with nature highlights the ongoing natural and cultural degradation as more than 'just another crisis'. This is because, without nature, humans will not be able to maintain their fundamental root of existence, identity, survival and sustenance.

Despite the urgency and interdependence with nature, our responses have been tailored to blindness with the application of anthropocentric techniques that had been employed by corporations to shape our perceptions of the natural world with intended motives of commercialization, profit, mass consumption and 'in-indoor' lifestyle as witnessed in the initial defiance of Kanai Dutt (in *The Hungry Tide*), Dinanath Datta and Tutul/Tipu (in *Gun Island*). In this context, nature appeared to be commodified, packaged and priced which had made the first nature "become second nature". This shift had alienated humans from their traditional eco-centric interconnectedness with nature which appeared to be elemental in the survival and sustenance of humans and the planet. Additionally, man's increased arrogance and ignorance had led him to believe in "a new and unprecedented power of penetration into the non-human world.... for a more sophisticated, accomplished dominance of it" (Pepper, 1995, 117, Maldonado, 2015, 24 and Purdy, 2015, 6). Such an exercise of reductionism had further restructured our ethical system with superficial rights, dignity and morality for unprivileged categories like women, children, non-westerners, animals and nature with a sense of *limited* love, reverence, fear and anxiety which eventually encouraged and justified human generated genocide. In fact, such a biased attitude had triggered several ecological mishaps in the contemporary period thereby increasing the severity and frequency of the ongoing environmental catastrophes in the Age of the Anthropocene. Towards this end, it has become important to understand the relationship between humans and nature to manage the contemporary environmental crisis that is gradually eroding the resilience and stability of the planet (Maldonado, 2015, 2 and 8). Keeping this in consideration, Adamson and Ghosh saw the Anthropocene as an "era of global warming" wherein "climate change itself is represented not so much as complex and "wicked" but as manageable, a challenge easily met if humans rely on their faculties of intelligence and creativity" (Ghosh, 2017, 35 and Adamson, 2013, 512).

In this regard, the impending apocalypse has become central to ecocriticism's projection of the environmental future which has shifted our "imagination to a sense of [an impending] crisis". Adding to this, Russell claimed that such an exercise would be "evocative rather than referential and performative rather than conclusive, thus radically undercutting any of the familiar thrills and consolations of an apocalyptic imaginary" (Buell, 1995, 285 and Russell, 2018, 216). Towards this human-induced environmental collapse, both Ghosh and Maldonado considered the limited role of humans as "destructive spoilers" who had crippled the social, national as well as environmental security (Maldonado, 2015, 23). This trend further led to the realization of humans as a geological force that had paved the way for Anthropocene with climate change as the greatest threat of the contemporary society. Keeping this in consideration, Ghosh had attempted to trace the question of human agency on a planetary level, both empirically as well as ethically. For this purpose, he addressed several ecological and climatological concerns and consideration in his works *The Hungry Tide* and *Gun Island*. This was done to acquire a holistic approach to the environmental crisis as witnessed in the postnatural period. In addition, his work *The Derangement* further illustrated on these issues by providing an extensive documentation of the social and environmental history of the ecosystem to trigger ethical considerations in our perceptions and imagination. In this effort, nature has been considered "as a victim of climate change, but a victim ready to fight back" which has eventually been represented as an "anthropomorphic monster" that is "too powerful, too grotesque, too dangerous... too accusatory", combative, vengeful and a "dangerous threat" to humankind and the planet (Ghosh, 2017, 43 and Adamson, 2013, 508-509). In this context, climate change has often been considered as a global phenomenon that has become one of the greatest threat to humanity as well as the ecosystem. Further, the ecological destabilization rising from climate change appears to threaten national security as witnessed through several social riots, migrations, civil disobedience and vandalism that were extensively addressed by Ghosh in both of his novels, *The Hungry Tide* and *Gun Island* while he also attempted to investigate the (un)manageable environmental crisis of the Anthropocene. This was because the drastic alteration of the ecosystem by invasive anthropogenic actions and calamities had threatened and eroded the permanence and stability of the planet which subsequently haunted our imagination and "intuition of apocalypse" (Ghosh, 2017, 167). Caught in this situation, there has been a steady rise of communal confusion, threat, chaos and anger which has furthered the denaturalization of the environment. Such a possibility was an outcome of indigenous tribes becoming exploiters, 'disrupters and destroyers' of the regional environment rather than continuing their previous role of eco-centric partners and stewards of the ecology. Following this, there have been unprecedented scale of disasters in regional, national and global levels as the stability, sanity and holism of the natural world have been largely manipulated and crippled. Keeping this in consideration, the Anthropocene appears to be a human-engineered phenomenon wherein humans are no longer simple biological agents. Instead they have evolved to become a significant geological force that has accelerated the mass extinction and biodiversity loss of the planet (Bates, 2018, 171, Hamilton, 2017, 15, Sandford, 2019, 18, Williston, 2015, 44, Sullivan, 2017, 25 and Schwagerl, 2014, 89).

     Through his ecological commitments, Ghosh attempted to highlight several social and environmental grievances that emerged through the postcolonial exercise of global marginalization of displaced social beneficiaries, indigenous communities and environmental species that have often been represented as a dire threat to the natural equilibrium of the local ecosystem. To this end, Ghosh considered the erratic climate change, global warming, unnatural migration and animal behaviors, increasing beaching and forest fires as immediate outcomes of invasive anthropogenic actions which he extensively addressed in his "cli-fi" sequel. Keeping to this, Gabriel and Garrard claimed the functioning of climate as "an idea of the imagination" that had been "profoundly embedded within our cultural traditions" thereby making it a natural as well as a social concept. Due to this, we often "find it more productive to 'see what climate change can do for us rather than what we seek to do, despairingly, for (or to) climate'" (2012, 120). Addressing the ecological crisis, Russell also stated that climate change as witnessed in the Anthropocene has largely been a human conditioned catastrophism that has altered the geological imprints of the planet. To this end, most of the literary, ecological and theoretical efforts have engaged with climate change with a sense of creative imagination and representation of the Apocalypse which has often been considered as a "master metaphor" that enabled the act of looking at a possible "brink of collapse" of humanity and the planet in the future. While this exercise pointed at our "cultural failure", it also attempted to ignite a trend of signaling timely warnings and awareness to help define, explore as well as resolve potential ecological concerns and considerations. In this regard, the initiative appeared crucial in our dire times as "the danger is not only imminent, but already well under way". Adding to this, Ghosh in *The Derangement* emphasized that while climate change may not appear to be an immediate "danger in itself; it is envisaged rather as a 'threat multiplier' that will deepen already existing divisions and lead to the intensification of a range of conflicts". In this sense, environmental catastrophe could be seen as a moment of social and ecological revelation wherein calamities attempt to rewrite "the destiny of the earth" with "urgent proximity of nonhuman presences" (Garrard, 2007, 93 and 95, Russell, 2018, 214 and Ghosh, 2017, 7, 9 and 192).

To this consideration, the increased occurrence of unnatural calamities in the Anthropocene had urged Ghosh to deal with changing imprints of nature as well as nonhumans' erratic behaviors and migration patterns that often resulted from the social and environmental risks, threats and insecurities. In fact, the planetary alterations have not only threatened the ecological equilibrium of the 'web of life' but also had accelerated the geological evolution which subsequently had strained the regenerative capacitates of the environment. This issue had been reported by Ghosh in his novels wherein he used a new emergent narrative style that applied science, facts, data, ideas, fiction and cultural myths in ecological literatures based on climate. Such an illustrious technique could be traced through the anti-vivisectionist cetologist, Piyali Roy who had come to the Sundarbans to scientifically study the dolphins. While engaging in this effort, Ghosh addressed the issue of megafauna extinction that had increased in its capacities with the endless human engineered alterations. To this end, he also highlighted the threat of multiple and massive episodes of marine beaching of crabs, fishes and dolphins which subsequently led to innumerable deaths in the marine ecosystem. This unnatural occurrence could be traced to the harmful industrial run-offs emitted by toxic fumes and wastes from the emerging refineries that have multiplied injudiciously in the Sundarbans. Subsequent to this, there had been a harmful contamination of the soil, air and water bodies in the archipelago. Along with these factors, an increased rate of uncontrolled deforestation has led to planetary and climatological problems like global warming and climate change which "influenced the [erratic] behavior and unnatural deaths of the dolphins" and other marine animals which thereafter contributed to the increased "stress on the mangrove forests, which act[ed] as a natural buffer to tropical cyclones and underpin the ecosystems for many marine invertebrate species and fish" (Trexler, 2014, 219).

In fact, the Sundarbans was not only exposed to the danger of climate change but also experienced social and environmental risks rising from the 'changing' river which posed as a threat with the constant rise of sea levels that had often led to unpredictable and aggressive flooding, tsunami, storms and salinization that jeopardized human life, settlements and daily routine. This furthered the tension over space, identity, security and survival among the social and ecological refugees of the region. In this context, the Anthropocene with its continual fear and hopelessness enabled the experience of a necessary catharsis following a monocultural representation of induced paranoia, conflict, anguish, violence and crisis. This complex psychological and emotional entanglement with natural calamities had been traced by Ghosh not only through Kusum and Kanai Dutt in *The Hungry Tide* as they encountered the man-eating tigers of the Sundarbans in their personal experiences but also through Dinanath Datta's unexpected exposure to scorpions and shipworms in Venice. In a similar manner, Tutul/Tipu's fits and Piyali Roy's nervous meltdown during the cyclone in Venice as mentioned in *Gun Island* reminded her of her past personal and ecological tragedy in the Sundarbans which had killed Tutul's father, Fokir in *The Hungry Tide*. In addition, the novel also recounted Horen's past experiences with cyclones which he compared as a "terrible battlefield massacre" that left behind innumerable "corpses everywhere and [while] the land was carpeted with dead fish and livestock" (Ghosh, 2013, 350). These episodes appear to a repressed response of PTSD as they re-experienced ecological trauma that reignited their level of stress and shocks which often led to impulsive crying and sweating that disabled them to continue their emotional stability and daily business (Morton, 2010, xxi). While these experiences may be considered insignificant on the surface level, Morton in his work, *Hyperobjects* considered such events in a fictional narration as a strategy "to awaken us from the dream that the world is about to end" and how "the end of the world is correlated with the Anthropocene, its global warming and subsequent drastic climate change, whose precise scope remains uncertain while its reality is verified beyond question" (2013, 7).

Conclusion

Through this exercise, there has been an effort to attach an 'additional' sense of deep commitment to ethics, stewardship and responsibilities of humans towards the environment as it has been exposed to human-oriented hybridization of the ecology. Such a tendency appears to be significant in the Anthropocene as it facilitated the opening and mobilization of traditional anthropocentrism through the concept of "new" anthropocentrism that has been addressed by several prominent scholars like Jerediah Purdy, Clive Hamilton, and Robert William Sandford. In this sense, "new" anthropocentrism appears to be "an extension of the older Enlightenment Project" (2015, 24). This realization has made us aware of nonhuman presences and functioning in literary, cultural and theoretical efforts that have always co-existed and co-evolved with human responses and existence. Such an understanding also opened possibilities for a shift from ecophobia among humans to a neutral 'interspecies' platform for ecological dialogues and deliberations as endorsed by ecocriticism and Anthropocene studies. In fact, such an effort could also provide effective direction to political and ethical re-engagement as well as re-consideration of anthropocentric beliefs. To this end, such an exercise appears crucial for the present generation living in the postnatural world of the Anthropocene which has witnessed a planetary "rupture" that has largely been human-engineered and whose adverse effects should be taken as a *timely warning* and visions to avoid the impending apocalypse (Hamilton, 2017, 41). To this end, several efforts have been undertaken by intellectual thinkers and environmentalists to achieve potential success in conservation and restoration of the ecosystem through awareness, activism, policies, and peaceful movements which have become more challenging in the Age of the Anthropocene.

References

Adamson, Joni. 2013. "Environmental Justice, Cosmopolitics, and Climate Change". In *The Cambridge Companion to Literature and the Environment*, edited by Louise Westling, 169-183.
Cambridge: Cambridge University Press.
Bates, Jonathan. 2018. "From 'Red' to 'Green'". In *The Green Studies Reader: From Romanticism to Ecocriticism*, edited by Laurence Coupe, 167-172. Oxon: Routledge.
Boddice, Rob. 2011. "Introduction: The End of Anthropocentrism". In *Anthropocentrism: Humans, Animals, Environments*, edited by Rob Boddice, 1-18. Leiden: Brill.
Buell, Lawrence. 1995.*The Environmental Imagination: Thoreau, Nature Writing, and the Formation of American Culture*. Cambridge: Belknap Press.
Chakrabarty, Dipesh. 2015. "The Climate of History: Four Theses". In *Ecocriticism: The Essential Reader*, edited by Ken Hiltner, 335-352. Oxon: Routledge.
Clark, Timothy. 2013. "Nature, Post Nature". In *The Cambridge Companion to Literature and the Environment*, edited by Louise Westling, 75-89. Cambridge: Cambridge University Press.
Deyo, Brain. 2018. "Tragedy, Ecophobia, and Animality in the Anthropocene". In *Affective Ecocriticism: Emotion, Embodiment, Environment*, edited by Kyle Bladow and Jennifer Ladino, 195-212. Lincoln: University of Nebraska Press.
Fromm, Harold. 1996. "From Transcendence to Obsolescence: A Route Map". In *The Ecocriticism Reader: Landmarks in Literary Ecology*, edited by Cheryll Glotfelty and Harold Fromm, 30-39. Athens: University of Georgia Press.
Gabriel, Hayden, and Garrard, Greg. 2012. "Reading and Writing Climate Change". In *Teaching Ecocriticism and Green Cultural Studies*, edited by Greg Garrard, 117-129. Hampshire: Palgrave Macmillan.
Garrard, Greg. 2007. *Ecocriticism: The New Critical Idiom*. Oxfordshire: Routledge.
Ghosh, Amitav. 2019. *Gun Island.* Gurgaon: Penguin Random House.
—. 2017. *The Great Derangement: Climate Change and the Unthinkable.* Gurgaon: Penguin Random House.

—. 2013. *The Hungry Tide.* Noida: HarperCollins.
Glotfelty, Cheryll. 1996. "Introduction: Literary Studies in an Age of Environmental Crisis". In *The Ecocriticism Reader: Landmarks in Literary Ecology,* edited by Cheryll Glotfelty and Harold Fromm, xv-xxxvii. Athens: University of Georgia Press.
Hamilton, Clive. 2017. *Defiant Earth: The Fate of Humans in the Anthropocene.* Crows Nest: Allen & Unwin, 2017.
Hamilton, Geoff, and Jones, Brain (eds). 2013. *Encyclopedia of the Environment in American Literature.* North Carolina: McFarland & Company.
Heise, Ursula K. 2008.*Sense of Place and Sense of Planet: The Environmental Imagination of the Global.* New York: Oxford University Press.
Hiltner, Ken. 2015. "Introduction". In *Ecocriticism: The Essential Reader,* edited by Ken Hiltner, xii-xvi. Oxon: Routledge.
Huggan, Graham, and Tiffin, Helen. 2010.*Postcolonial Ecocriticism: Literature, Animals, Environment.* Oxon: Routledge.
Kluwick, Ursula. 2014. "Talking about Climate Change: The Ecological Crisis and Narrative Form". In *The Oxford Handbook of Ecocriticism,* edited by Greg Garrard, 502-516. New York: Oxford University Press.
Krebber, Andre. 2011. "Anthropocentrism and Reason in *Dialectic of Enlightenment*: Environmental Crisis and Animal Subject". In *Anthropocentrism: Humans, Animals, Environments,* edited by Rob Boddice, 321-339. Leiden: Brill.
Maldonado, Manuel Arias. 2015.*Environment and Society: Socionatural Relations in the Anthropocene.* Cham: Springer.
Manes, Christopher. 1996. "Nature and Silence". *The Ecocriticism Reader: Landmarks in Literary Ecology,* edited by Cheryll Glotfelty and Harold Fromm, 15-29. Athens: University of Georgia Press.
McKibben, Bill. 2003.*The End of Nature.* London: Bloomsbury.
Morton, Timothy. 2010. *The Ecological Thought.* Cambridge: Harvard University Press.
--- 2013. *Hyperobjects: Philosophy and Ecology after the End of the World.* Minneapolis: University of Minnesota Press.
Nimmo, Richie. 2011. "The Making of the Human: Anthropocentrism in Modern Social Thought". In *Anthropocentrism: Humans, Animals, Environments,* edited by Rob Boddice, 59-80. Leiden: Brill.
Pepper, David. 1995.*Eco-Socialism: From deep ecology to social justice.* London: Routledge.
Purdy, Jedediah. 2015. *After Nature: A Politics for the Anthropocene.* Cambridge: Harvard University Press.

Russell, Allyse Knox. 2018. "Futurity without Optimism: Detaching from Anthropocentrism and Grieving Our Fathers in *Beasts of the Southern Wild*". In *Affective Ecocriticism: Emotion, Embodiment, Environment*, edited by Kyle Bladow and Jennifer Ladino, 213-232. Lincoln: University of Nebraska Press.

Sandford, Robert W. 2019. *The Anthropocene Disruption*. Alberta: Rocky Mountain Books.

Scranton, Roy. 2015. *Learning to Die in the Anthropocene: Reflections on the End of a Civilization.* San Francisco: City Light Books.

Schwägerl, Christian. 2014. *The Anthropocene: The Human Era and How It Shapes Our Planet*. Translated by Lucy Renner Jones. London: Synergetic Press.

Singer, Peter. 2011.*Practical Ethics*. New York: Cambridge University Press.

Soper, Kate. 2000. "The Idea of Nature". *The Green Studies Reader: From Romanticism to Ecocriticism*, edited by Lawrence Coupe, 123-126. Oxon: Routledge.

Sullivan, Heather I. 2017. "The Dark Pastoral: A Trope for the Anthropocene". In *German Ecocriticism in the Anthropocene*, edited by Caroline Schaumann and Heather I. Sullivan, 25-44. New York: Palgrave Macmillan.

Trexler, Adam. 2014. "Mediating Climate Change: Ecocriticism, Science Studies and *The Hungry Tide*". In *The Oxford Handbook of Ecocriticism*, edited by Greg Garrard, 205-224. New York: Oxford University Press.

—. 2015.*Anthropocene Fictions: The Novel in A Time of Climate Change*. Charlottesville: University of Virginia Press.

Williston, Byron. 2015.*The Anthropocene Project: Virtue in the Age of Climate Change*. Oxford: Oxford University Press.

CHAPTER NINE

THE ECOLOGICAL CONSCIENCE OF IAN MCEWAN'S FICTION

ANASTASIA LOGOTHETI

Introduction

One of the achievements of ecocriticism has been to redirect critical attention to the significance of place and the need to explore the environmental consciousness of the literary text. The environmentally sensitive side of the contemporary English novel has already been noted in seminal ecocritical studies by Lawrence Buell, Greg Garrard, and Lawrence Coupe.[1] In *The Environmental Imagination* Lawrence Buell notes the "anti-environmental" tendencies of traditional practitioners of literary criticism;[2] conversely, Greg Garrard explains, "ecocritics"[3] follow "ecophilosophers in identifying anthropocentrism as the core conceptual problem with Western civilization in its relations with more-than-human nature."[4] In an essay provocatively entitled "Ecocriticism: What Is It Good For?" Robert Kern suggests that ecocriticism "becomes most interesting" when it helps us "recover the environmental character" of the literary text.[5] Foregrounding the environmental concerns inherent in the relationship between the Earth and its inhabitants, the "modern novel, with its emphasis on private feeling

---

[1] Buell, *The Environmental Imagination* (1995), and *The Future of Environmental Criticism* (2005); Gerrard, *Ecocriticism* (2004); Coupe, *The Green Studies Reader* (2000).
[2] Buell, *The Environmental Imagination*, 85.
[3] Garrard (in "Solar," 93) defines ecocritics as "literary critics with an environmental orientation."
[4] Garrard, "Reading," 224.
[5] Kern, "Ecocriticism," 260.

as the source of public action" can be, in the words of Dominic Head, "an appropriate vehicle for a Green agenda."[6]

The connection between human actions and the environment as well as the profound implications that the local bears on the global are issues of concern in the work of Ian McEwan. A celebrated contemporary novelist who combines popularity with critical acclaim, McEwan has been devoting ever increasing attention to ecosystem problems in his work. As Garrard notes, McEwan's fiction combines "good science" with "informed skepticism."[7] Investigating the repercussions of alienation from the self and the natural environment, McEwan revealed his interest in environmental debates from the beginning of his career. From the early environmental dystopias of *The Cement Garden* (1978) and *The Child in Time* (1987) to the satire of *The Cockroach* (2019) and the counterfactual past of *Machines Like Me* (2019), McEwan has been drawing attention to the most critical dilemma of our time: the struggle between anthropocentrism and environmental ethics.

This chapter aims to reveal McEwan's deep commitment to ecology through the discussion of two of his works published in the first decade of the twenty-first century: *On Chesil Beach* (2007) and *Solar* (2010) demonstrate not only an environmental consciousness, that is, an unmistakable sense of place which provides the narrative with deep roots into the physical world outside the text, but also an awareness of the repercussions of the alienation of humankind from the natural world. As a result, McEwan's novels explore the repercussions of the lack of an "ecological conscience," a term which refers to the "ethics of community, that is, the ethics of living in accord with the welfare of the ecological community."[8]

On Chesil Beach (2007)

Set in mid-July 1962 at the eponymous pebbly shore, McEwan's tenth novel, *On Chesil Beach*, has not attracted the attention of ecocritics despite the use of a unique natural setting to highlight the protagonists' profound sense of alienation within a context of crisis which includes the potential for change, both personal and national. In the first four chapters of this novella-length narrative newly-weds Edward and Florence attempt

---

[6] Head, "Ecocriticism and the Novel," 240
[7] Garrard, "Reading," 223.
[8] I am borrowing the term "ecological conscience" from Leopold Aldo who coined it (according to the "Glossary of Selected Terms" in Buell, *The Future of Environmental Criticism*, 140).

sexual intercourse on the four-poster bed of the honeymoon suite at a Georgian inn on the Dorset coast. In the final chapter the action transfers to the beach, which divides "with its infinite shingle" the Fleet Lagoon from the English Channel.[9] As the preposition in the novel's title suggests, the climax occurs on a strip of land which forms a natural border, signifying not only the division between the lovers but also marking the difference between 1962 and the rest of "that famous decade."[10]

Focusing on the events of a disastrous wedding night, the novel exposes the unwillingness of a young couple to act in a manner that would encourage safeguarding of each other and of the world although they claim that they care for the environment. The young protagonists meet at a "a lunchtime meeting of the local CND," the Campaign for Nuclear Disarmament, which Edward reluctantly attends out of a vague sense of "duty to save the world."[11] Edward and Florence are prisoners of their times and share its unease: they are paradigmatic of a regression which constitutes the undercurrent of the wave which ushered in the Swinging Sixties. Although they represent a new generation accepting of the post-colonial status of England as a "minor power,"[12] the protagonists are not immune to a defeatist attitude which permeates society and mirrors their own emotional inertia. Their marriage at the age of twenty-two is an escape from their families and a socially approved remedy to the insignificance of their youth. The couple's inexperience in sexual matters compounded by their unwillingness to talk about intimacy and desire makes their first night fraught with challenge.

Despite a courtship which lasts several months, based on mutual admiration of minds and the recognition of their outsider status in their respective family environments, Edward and Florence have not found a way to communicate to each other their innermost fears: "Nothing was ever discussed—nor did they feel the lack of intimate talk."[13] Concerned more about duty and decorum than revealing their anxieties, the protagonists do not bare their naked selves literally or metaphorically until their wedding night. Following Edward's premature ejaculation, Florence abandons groom and hotel bed to seek comfort in nature after what has been for her a traumatic sexual encounter: she runs to the "warm and moist" beach to

---

[9] Ian McEwan, *On Chesil Beach*, 3-4. All subsequent references to this novel are to the 2008 Vintage edition.
[10] Ibid, 6.
[11] Ibid, 47.
[12] Ibid, 24.
[13] Ibid, 21.

"escape herself."[14] Eventually, she rests against the "smoothed and hardened" trunk of "a great fallen tree" which the elements have picked clean.[15] The narrative voice suggests that the protagonist tries to receive from nature the physical comfort of an infant cradled in maternal arms. Both romantic child and Eve hiding in shame and regret in a postlapsarian Edenic setting, Florence seeks comfort through therapeutic "nestling" against a washed-up tree trunk. However, the "darkening shingle" of this famous coast is not a welcoming place for Florence.[16] She has cut herself off from the environment and cannot reconnect into the natural world or herself.

Despite listening to the waves and the song of a distant blackbird, despite breathing in the fresh sea air, the protagonists are unable to connect to the natural surroundings and therefore to one another. On the "famous shingle spit,"[17] the couple have one final hurtful verbal exchange: "what they had here, on the shores of the English Channel," the narrator explains, "was only a minor theme in the larger pattern."[18] Equally numb to the environment and to themselves, they exist in time but not in place. Insulted by Florence's rejection of any further sexual activity and her proposal of an open marriage, which he terms "ridiculous" and "an insult," Edward rejects any compromise, accusing Florence of acting as if it were 1862.[19] Their inability to communicate in any other way except through words imprisons them as much as walking on the shingles along the long shore creates the illusion that they are immobile.

The novel's denouement toys with the expectations of readers for some unexpected reversal of fortune, but the protagonists never meet again. After watching Florence "hurry along the shore, the sound of her difficult progress lost to the breaking of small waves,"[20] Edward remains on the beach to indulge in "a day-dream of self-pity" while "hurling stones at the sea and shouting obscenities."[21] Nature affords him the complete freedom he sought in his short-lived marriage to Florence but locked as he is inside himself, he can derive no pleasure as he walks "up and down on the exhausting shingle."[22] The tree trunk, against which he slumps as Florence had done earlier, cannot restore his spirit or rescue him from a self-indulgent

---

[14] Ibid, 140.
[15] Ibid, 141.
[16] Ibid, 141.
[17] Ibid, 142.
[18] Ibid, 146.
[19] Ibid, 156.
[20] Ibid, 166.
[21] Ibid, 158.
[22] McEwan, *On Chesil Beach*, 158.

egotism. As Edward realizes, "we could be in paradise. Instead, we're in this mess."[23]

Educated and rational adults, who are alienated from themselves and from others in a post-industrial world fraught with hints of dystopia, McEwan's characters are focused on themselves and unable to connect with each other. They lack a sense of self as members of a community because they lack an ecological conscience. Through this condensed narrative McEwan engages with the politics of post-World-War-II Britain, focusing on the country's diminished global importance and its outdated class hierarchies, exposing a cultural context of entrenchment, loss, and lives lived under the threat of nuclear war and away from nature.

Solar (2010)

A satire exploring contemporary debates about climate change, *Solar* is the novel McEwan published after *On Chesil Beach*. Ecocritics eagerly awaited the publication of *Solar* which was "hailed as the first novel on climate crisis by a major British author."[24] Even before the novel's publication Greg Garrard, in an article entitled "Ian McEwan's Next Novel and the Future of Ecocriticism," predicted the significance of McEwan's forthcoming work for ecocritics. Once the novel was published, Garrard was disappointed that *Solar* was "not apocalyptic."[25] Kerridge found the novel's combination of the "picaresque tradition" and "mock-heroic comedy" with "Hardyesque irony" and "fatalism" to be reductive; he notes that the novel lacks the "full emotional and moral range of which the realistic novel is capable."[26]

Thus, the genre McEwan chose for *Solar* surprised and disappointed ecocritics who expected a dystopian or apocalyptic treatment of the topic.[27] While some scholars were disappointed that the novel "does not aim to stimulate activism,"[28] others found the "comic representation of the world of science" to be a "strategy aimed at debunking the idea of social advancement."[29] As Hsu notes, "a space for ethics" exists in *Solar*[30] while Marzec finds that through a protagonist who is "comically self-destructive

---

[23] Ibid, 150.
[24] Bracke, "Contemporary," 431.
[25] Garrard, "*Solar*: Apocalypse Not," 100.
[26] Kerridge, "The Single Source," 157-59.
[27] Garrard, "*Solar*: Apocalypse Not," 93.
[28] Berndt, "Science as Comedy," 87.
[29] Ibid, 86.
[30] Hsu, "Truth," 326.

and blithely unaware of his own extreme threat to ecosystems", the novel castigates "the kind of siloed logic that fails to consider the complexity of ecological relations continually violated by neoliberalism."[31] Although Garrard argues that "McEwan's commitment to Enlightenment values makes fundamental critique of capitalism almost inconceivable to him,"[32] Traub suggests that "cli-fi humor" is a means of exposing "capitalist greed."[33] Moreover, Bracke notes that the "pluriformity of views on nature" in the novel offers an opportunity for ecocriticism to adopt a "more skeptical attitude towards its own assumptions."[34] As Traub notes, "comic strategies" can be considered as "alternative means by which environmental risks might be represented and ethically engaged."[35] Similarly, Zemanek finds *Solar* to be a "satiric-allegorical risk narrative" which can be classified as "a new form of eco-fiction"[36] with "greater potential to incite reflection."[37]

Solar is the product of McEwan's extensive research into the science of climate change but also draws on the author's own experiences which allow him to connect the environmental crisis to "human nature."[38] In the Acknowledgements page, which is published at the end of *Solar*, McEwan specifies the various areas of research he undertook as well as his collaboration with science experts in relation to theoretical physics and the challenges of climate change. From the Cape Farewell expedition to the Arctic in March 2005, to the Potsdam conference in October 2005, to visiting solar energy facilities in Golden, Colorado, and living in New Mexico, McEwan experienced first-hand the science debates and landscapes that inform the novel.[39]

---

[31] Marzec, "Contemporary Anthropocene Novels," 338.
[32] Garrard, "*Solar*: Apocalypse Not," 97.
[33] Traub, "From the Grotesque," 103.
[34] Bracke, "Contemporary," 433-4.
[35] Traub, "From the Grotesque," 104.
[36] Zemanek, "A Dirty Hero's Fight," 52.
[37] Ibid, 59.
[38] Anthony, "Ian McEwan."
[39] McEwan participated in Cape Farewell's 2005 art / science expedition that took place onboard the *Noorderlicht*, which was locked in ice at Tempelfjorden, Spitsbergen, just north of the 79th parallel (https://capefarewell.com/2005.html). For six days twenty artists, scientists and journalists experienced the arctic environment in extreme temperatures of -30°C. McEwan's experience is described in the essay "A Boot Room in The Frozen North." In March 2007 McEwan joined Professor H J Schellnhuber, Germany's Chief Government Advisor on Climate, in conversation at the Bucerius Law School, Hamburg. The event was presented alongside the Cape Farewell exhibition Art & Climate Change (https://capefarewell.com/who-we-

As the title suggests, *Solar* relates to the greatest environmental challenge of the twenty-first century, global warming, and references one possible solution, solar energy. In 2009, responding to the question "what will change everything", McEwan responds, "I hope I live to see the full flourishing of solar technology."[40] While in interviews the author declares that he is "optimistic" about finding a solution to climate change, he emphasizes the connection between the climate and global poverty[41] and notes that change will come through science. Still, Michael Beard, the theoretical physicist who is the protagonist of *Solar*, is, in McEwan's own words, "an intellectual thief. ... sexually predatory. ... a compulsive eater, a round and tubby fellow who has profound self-belief. ... rackety, quarrelsome, competitive, greedy, ambitious, politicking."[42]

Fifty-three-year-old Michael Beard, a Nobel laureate who heads the National Centre for Renewable Energy near Reading in England, exemplifies all the vices that have brought the globe to a critical point. Beard may be intelligent but without empathy or morality his life dissolves into chaos: he neglects his health beyond repair, he always acts out of self-interest and he does not consider the long-term effects of his actions and decisions. Despite the potential for developing an environmental consciousness through the opportunities provided throughout the novel, Beard misses every chance and does not acquire an ecological conscience.

Divided in three parts, the novel is set during the first decade of the twenty-first century. Part One of the novel is set in the year 2000 when "the earth's fate hung in the balance."[43] Michael Beard no longer has any "new ideas" but gives the "same series of lectures" on the "Beard-Einstein Conflation" for which he was awarded the Nobel prize for Physics decades earlier.[44] Beard considers that his talent is "tiny" compared to the "genius" of Einstein[45] but he enjoys recognition; he also earns considerable income through his university salary, media appearances, and lecture fees.[46] As head

---

are/creatives/63-ian-mcewan.html). In Oct 2007 McEwan participated in the Potsdam Symposium on Global Sustainability (http://www.nobel-cause.de/potsdam-2007) again at the invitation of Prof Schellnhuber.
[40] McEwan, "The Full Flourishing."
[41] Roberts, "A Thing," 190.
[42] Zalewski, "The Background."
[43] Ian McEwan, *Solar*, 218. All subsequent references to this novel are to the 2010 Jonathan Cape edition.
[44] Ibid, 14-5.
[45] Ibid, 50.
[46] Ibid, 15-6.

of the newly set up National Centre for Renewable Energy, Beard is promoting the creation of a "Wind turbine for Urban Domestic Use (WUDU)."[47] The project is costly and impossible to realize but Beard does not want to risk his reputation by acknowledging this failure;[48] instead, he seeks to increase government funding for WUDU through media attention. From the beginning the novel exposes the protagonist's selfishness as indicative of the lack of an environmental consciousness.

One of the young post-doc researchers at the Centre, Tom Aldous, is interested in solar energy instead of WUDU; he proposes developing "artificial photosynthesis" in response to climate change.[49] While Aldous keeps insisting on the urgency of the environmental crisis and the need for "a new energy source for the whole of civilization,"[50] thus revealing his ecological conscience, Beard does not believe any of the "predictions:" he is uninterested in anything idealistic or apocalyptic.[51] Even after participating in an expedition dedicated to climate change for which he travels to the island of Spitsbergen in the Svalbard archipelago in northern Norway, Beard is convinced that the planet cannot be saved by "science, or art, or idealism."[52] The protagonist's comic adventures in a frozen landscape near the North Pole constitute a satirical piece of thirty pages which further highlights the character's moral flaws.[53] Beard lacks motivation: he is nihilistic and believes only in "slow inner and outer decay."[54] He recognizes that he should "lose weight, get fit, live in a simple, organised style," that he should "act now or die early," but these resolutions remain unrealized.[55]

At the age of fifty-three and married for the fifth time, Beard is consumed, throughout the first part of the novel, by his 34-year-old wife's affairs, first with a builder, Rodney Tarpin, and then with Tom Aldous. Primary-school-teacher Patrice is unfaithful in retaliation for her husband's infidelities: Beard has managed "eleven affairs in five years."[56] Initially, Beard is determined to win her back so he confronts Tarpin about giving Patrice a black eye; Rodney becomes violent and slaps Beard. When Patrice rejects Tarpin, the latter threatens more violence through letters and phone

---

[47] Ibid, 23.
[48] Ibid, 28.
[49] Ibid, 25.
[50] Ibid, 34.
[51] Ibid, 75-7.
[52] McEwan, *Solar*, 80.
[53] Ibid, 49-80.
[54] Ibid, 66.
[55] Ibid, 72-3.
[56] Ibid, 47.

calls. In the meantime, Patrice has taken Aldous as a lover, which Beard discovers when he returns from the journey to the Arctic. Although he has already decided on separating from his wife and never marrying again, Beard assumes the part of the "cuckold" when he finds the young man living in the marital home.[57] A confrontation between husband and lover, the established scientist and the post-doc "genius,"[58] ends with the young man's accidental fall and fatal injury. Beard makes use of this "improbable" domestic accident and frames Tarpin for Aldous's murder.[59] At the end of Part One Tarpin receives a life sentence while Beard is newly divorced and in possession of Aldous' pioneering work on artificial photosynthesis.[60]

Part Two of the novel is set in 2005, the year the Kyoto Protocol[61] came into force: Beard is fifty-eight, "thirty-five pounds overweight,"[62] and living in a "neglected" apartment in Marylebone,[63] having moved away from the Belsize-Park residence of his disastrous fifth marriage. He is now pursuing Aldous's research project on solar energy but not because he has grown an environmental consciousness. From denying climate change in Part One, in Part Two the protagonist uses the environmental crisis to promote his own interests and to make profit. Having started his own company and found a business partner, Beard seeks to convince "institutional investors and pension-fund managers"[64] to invest in "affordable clean energy."[65] Ironically, the protagonist uses not only the ideas but also the arguments of his late rival to persuade investors. At a Savoy-hotel gathering Beard insists that "the problem lies … in ourselves, our own follies and unexamined assumptions" and that "the acquisition of new information forces us to make a fundamental reinterpretation of our situation."[66] While he delivers his speech movingly, he is incapable of following this advice himself: before addressing the room Beard "bolted down" nine salmon

---

[57] Ibid, 84.
[58] Ibid, 83.
[59] Ibid, 90.
[60] Ibid, 103.
[61] According to the United Nations Framework Convention on Climate Change (https://unfccc.int/kyoto_protocol), the entity tasked with supporting the global response to the threat of climate change, the Kyoto Protocol was adopted in 1997 and entered into force in 2005, committing industrialized countries to limit and reduce greenhouse gases (GHG) emissions in accordance with agreed individual targets.
[62] McEwan, *Solar*, 118.
[63] Ibid, 109.
[64] Ibid, 112.
[65] Ibid, 150.
[66] Ibid, 155.

sandwiches,[67] which he vomits undigested after his speech, exemplifying his inability to reflect on his own situation, even to control his basic urges.

Similarly, Beard causes an intense but brief media scandal which results from expressing an unpopular view about "an evolutionary past bearing down in some degree on cognition, on men and women, on culture"[68] in response to a journalist's query about gender inequality in the sciences and why "women" are "underrepresented in physics."[69] Despite the loss of a few academic posts and professional connections, which are soon replaced by others, the protagonist's standing is not diminished. Although Beard recalls such experiences as anecdotes, he does not examine or acknowledge his own responsibility as public speaker or as representative of science; his goals are selfish and his desires ephemeral. Beard is "self-sufficient, self-absorbed, his mind a cluster of appetites and dreamy thoughts:"[70] through such commentary the text reminds readers frequently of the effects of lacking an ecological conscience.

In Part Two the protagonist considers himself "no crueller" than most people although also aware he is "sometimes greedy, selfish, calculating."[71] His current relationship with forty-year-old, dance-clothes shop-owner Melissa Browne, who is "kind," "patient," and "artless" but also longing to be a mother, occupies him less than "his cranky affair with sunbeams."[72] When Melissa reveals her pregnancy, the protagonist thinks of all the "terminations" he has persuaded women to have in the last forty years.[73] He is certain that he can persuade Melissa as well that "this child could not be."[74] Though "never a complete cad,"[75] Beard does not want either to separate or to commit to this relationship. The novel offers Beard considerable opportunities for self-improvement and for using his influence to contribute positively to the science of climate change; yet, he fails to connect to ecology and to community.

While Beard "comfortably shared all of humanity's faults,"[76] he is determined not to be "too hard on himself."[77] Despite his many health

---

[67] Ibid, 146-7.
[68] Ibid, 144.
[69] McEwan, *Solar*, 133.
[70] Ibid, 169.
[71] Ibid, 170.
[72] Ibid, 110-21.
[73] Ibid, 172.
[74] Ibid, 185.
[75] Ibid, 159.
[76] Ibid, 171.
[77] Ibid, 186.

issues, he does not wish to see a doctor, "make a full confession" and "hear himself condemned."[78] Beard objects to being called "a plagiariser" even if he seeks "sole attribution" for the innovative ideas he has stolen from Tom Aldous.[79] Beard oscillates between contradictory attitudes, from "let there be light!" to "let the planet go to hell," which show him to be "a monster of insincerity."[80] At the end of Part Two Beard finds himself a reluctant father-to-be with a "purplish blotch on the back of his hand."[81] Still, the protagonist prefers to delude himself: he claims that he is doing the "hard work" on solar energy that would make the world "a better place."[82]

Part Three, the final part of the novel, is set in 2009, the year of the Copenhagen Accord:[83] Beard is sixty-two, in poor health and "carrying an extra sixty-five pounds."[84] He is an absent father to a three-year-old daughter, Catriona, "an intimate, sociable girl of near unbearable sensitivity," who loves her "Daddy" and does not notice his "reticence."[85] Unlike Melissa, Beard has not been "transformed" by paternity; his life has deteriorated along familiar patterns.[86] Nonetheless, he is experiencing an "unexpected sexual renaissance" with Darlene, a fifty-one-year-old waitress, he has met in New Mexico.[87] Having continued to pursue the solar-energy project, Beard owns seventeen patents and is determined to "bring to the world artificial photosynthesis on an industrial scale."[88] After much effort, delays and setbacks, a test site, the "Lordsburg Artificial Photosynthesis Plant," has been established in New Mexico,[89] and a demonstration is scheduled to take place attended by scientists, entrepreneurs, and the press. Beard is "hungry for public triumph" and ready to complete the "eight-year

---

[78] Ibid, 185.
[79] Ibid, 186-7.
[80] Ibid, 144; 183; 171.
[81] Ibid, 185.
[82] Ibid, 186.
[83] When asked for his assessment of the Copenhagen Accord, Fuqiang Yang, director of global climate solutions at WWF International and one of the "leading climate change experts," notes that the Copenhagen Climate Change Conference failed to produce "a legally binding treaty to reduce greenhouse gas emissions" (*Guardian*, 22 December 2009, https://www.theguardian.com/environment/2009/dec/22/copenhagen-climate-deal-expert-view).
[84] McEwan, *Solar*, 239.
[85] Ibid, 220-1.
[86] Ibid, 225.
[87] Ibid, 246.
[88] Ibid, 205.
[89] Ibid, 244.

journey from the slow deciphering of the Aldous file to lab work, refinements, breakthroughs, drawings, filed tests."[90]

Yet, past transgressions result in disaster: not only Beard's health issues are worsening—the blotch on his hand is diagnosed as melanoma—but also Tarpin, released from prison on good behaviour, travels to America to find Beard. Similarly, a lawyer authorized by the National Centre for Renewable Energy where Aldous worked is travelling to the desert to meet him. Tarpin has come to ask for a job: although Beard realizes the former strongman, now "delusional" and "a fantasizing fool," is someone who "probably deserved a break," he denies Tarpin's request.[91] Following the confrontation with the resentful ex-convict, Beard learns that Melissa and Catriona are also coming to Lordsburg to claim him since Darlene has informed them that she and Beard "are getting married."[92] Even when Beard insists that he is "not marrying anyone", Darlene remains unthwarted.[93]

The next meeting, with the lawyer representing the Centre, convinces Beard's business partner, Toby Hammer, to distance himself from the project and take legal action to ensure that the fast-increasing debt of three and a half million dollars will fall on Beard alone. The lawyer presents evidence which overwhelmingly proves that the patents Beard holds are based on Aldous's ideas and the British government "are keen to see the Centre own the patents and show the taxpayer a decent return."[94] Even more catastrophically, "someone's taken a sledgehammer to the panels" and all the equipment to be used in the demonstration is "shattered;" Beard guesses that Tarpin must have taken his revenge. When Hammer calls him a liar and a thief who "deserves everything that's coming,"[95] Beard does not protest: he realizes how he has contributed to his downfall, even if he has not grown a conscience, ecological or otherwise.

On the final page of the novel, Beard is trying to come to terms with the repercussions of the various long-term deceptions he has harbored about himself and with the fact that he will not "be saving the world;" he has little time to contemplate the lies his inflated ego has been nurturing instead of developing communal connections that would enable an environmental conscience. When Darlene as well as Melissa come to find him, they appear, like the natural world he has betrayed, "stormy" and

---

[90] McEwan, *Solar*, 240.
[91] Ibid, 258-60.
[92] Ibid, 261.
[93] Ibid, 265.
[94] Ibid, 272.
[95] Ibid, 277.

"furious."[96] As the protagonist greets his daughter, Catriona, he "felt in his heart an unfamiliar, swelling sensation."[97] Although left unexplained, the phrase suggests that the "growing risk of congestive heart failure," of which the doctor warned Beard after his most recent check-up,[98] may be the cause of this sensation.

Solar is a relentless indictment of the abuse of the natural world and of the self-centred indifference that may lead to collective ruin. The novel ends with the protagonist suspended between life and death, public and private disgrace. Whether the reader believes that Michael Beard survives or dies, no doubt exists about his status: he is a failed father, a deceitful partner, and a discredited scientist. Although intelligent and not unaware of his indulgences and self-deceptions, Beard is too selfish to feel properly guilty and to embrace change. Michael Beard, the gluttonous, deceitful, immoral scientist, who is as flawed a protagonist as McEwan has ever created, exemplifies the themes this tragicomic novel connects with the climate crisis. The lack of any kind of conscience, ecological or otherwise, is presented as the cause of this profound disaster, personal, institutional, and scientific.

Conclusion

As the discussion of these two novels demonstrates, the environmental crisis is a subject with which Ian McEwan has engaged consistently. The disastrous alienation from the natural world constitutes the common ground upon which McEwan's fiction foregrounds the necessity of an ecological conscience. "The planet does not turn for us alone" declares another couple in the pacifist oratorio *Or Shall We Die?* (1983), of which McEwan wrote the libretto and which concerns the threat of nuclear war. "Are we too late to save ourselves?" they wonder: "Shall we change, or shall we die?" they enquire repeatedly. As McEwan notes in the "Programme Notes," "this oratorio grew out of the conviction that the responsibility of the survival of our species is not limited to governments, but is collective, involving every single one of us. It is as if we had been set a simple test of maturity; we either pass it or perish."[99]

Through the plight of characters in *Solar* and *On Chesil Beach*, the novels comment unfavourably on the lack of connection to nature and the

---

[96] Ibid, 278.
[97] Ibid, 279.
[98] Ibid, 239.
[99] Ian McEwan, "Programme Notes."

absence of environmental consciousness. Castigating the exploitation of people and ideas for profit as well as the lack of concern for shared benefit, McEwan's fiction foregrounds the need for honesty, insists on community ethics, and demonstrates commitment to an ecological conscience.

References

Anthony, Andrew. "Ian McEwan: The Literary Novelist with a Popular Appeal." *Observer*, 28 February 2010.
Buell, Lawrence. *The Environmental Imagination: Thoreau, Nature Writing, and the Formation of American Culture*. Cambridge, MA: Harvard University Press, 1995.
Buell, Lawrence. *The Future of Environmental Criticism: Environmental Crisis and Literary Imagination*. Malden, MA: Blackwell, 2005.
Berndt, Katrin. "Science as Comedy and the Myth of Progress in Ian McEwan's *Solar*." *Mosaic* 50, no. 4 (2017): 85-101.
Bracke, Astrid. "The Contemporary English Novel and Its Challenges to Ecocriticism." In *The Oxford Handbook of Ecocriticism*, edited by Greg Garrard, 423-39. Oxford: Oxford University Press, 2014.
Brown, Tina. "Ian McEwan on *Solar*'s 'Fat, Greedy' Protagonist" in "The Beast Bar: Tina Brown Talks to Ian McEwan." *The Daily Beast*. 14 April 2010. https://youtu.be/m90XC1ZNYGg.
Coupe, Lawrence, ed. *The Green Studies Reader: From Romanticism to Ecocriticism*. London: Routledge, 2000.
Garrard, Greg. *Ecocriticism*. London: Routledge, 2004.
Garrard, Greg. "Ian McEwan's Next Novel and the Future of Ecocriticism." *Contemporary Literature* 50, no. 4 (2009): 695-720.
Garrard, Greg. "Reading as an Animal: Ecocriticism and Darwinism in Margaret Atwood and Ian McEwan." In *Local Natures, Global Responsibilities: Ecocritical Perspectives on the New English Literatures*, edited by Laurenz Volkmann, Nancy Grimm, Ines Detmers, and Katrin Thomson, 223-42. Amsterdam: Rodopi, 2010.
Garrard, Greg. "*Solar*: Apocalypse Not." In *Ian McEwan: Contemporary Critical Perspectives*, second edition, edited by Sebastian Groes, 93-101. London: Bloomsbury Academic, 2013.
Head, Dominic. "Ecocriticism and the Novel." In *The Green Studies Reader: From Romanticism to Ecocriticism*, edited by Lawrence Coupe, 235-41. London: Routledge, 2000.
Hsu, Shou-Nan. "Truth, Care, and Action: An Ethics of Peaceful Coexistence in Ian McEwan's *Solar*." *Papers on Language and Literature* 52.4 (2016): 326–49.

Kern, Robert. "Ecocriticism: What Is It Good For?" In *The ISLE Reader: Ecocriticism 1993-2003*, edited by Michael P. Branch and Scott Slovic, 258-281. Athens, GA: University of Georgia Press, 2003.

Kerridge, Richard. "The Single Source." *Ecozon@: European Journal of Literature, Culture and Environment* 1, no. 1 (2010): 155–61. doi: https://doi.org/10.37536/ ECOZONA.2010.1.1.334.

Marzec, Robert P. "Contemporary Anthropocene Novels: Ian McEwan's *Solar*, Jeanette Winterson's *The Stone Gods*, Margaret Atwood's *Oryx and Crake* and *The Year of the Flood*." In *New Directions in Philosophy and Literature*, ed. David Rudrum, Ridvan Askin and Frida Beckman, 338-60. Edinburgh: Edinburgh University Press, 2019.

McEwan, Ian. "A Boot Room in the Frozen North." Cape Farewell 2005 Art/Science Expedition. https://capefarewell.com/explore/215-a-boot-room-in-the-frozen-north.html. Reprinted in the *Guardian* (19 March 2005) as "Save the Boot Room, Save the Earth." https://www.theguardian.com/artanddesign/2005/mar/19/art1.

McEwan, Ian. *The Cement Garden*. London: Jonathan Cape, 1978.

McEwan, Ian. *The Child in Time*. London: Jonathan Cape, 1987.

McEwan, Ian. *The Cockroach*. London: Jonathan Cape, 2019.

McEwan, Ian. "The Full Flourishing of Solar Technology." Response to Annual Question 2009 on Edge.org. https://www.edge.org/response-detail/10321.

McEwan, Ian. *Machines Like Me*. London: Jonathan Cape, 2019.

McEwan, Ian. *On Chesil Beach*. London: Vintage, 2008.

McEwan, Ian. "Programme Notes" for *Or Shall We Die? (1983)*. In *A Move Abroad: Or Shall We Die? and The Ploughman's Lunch*. London: Picador, 1989. https://global.oup.com/academic/product/or-shall-we-die-9780193354029?cc=gr&lang=en&#.

McEwan, Ian. *Solar*. London: Jonathan Cape, 2010.

Roberts, Ryan. "'A Thing One Does:' A Conversation with Ian McEwan." In *Conversations with Ian McEwan*, edited by Ryan Roberts, 188-201. Jackson, MS: University Press of Mississippi, 2010.

Traub, Courtney. "From the Grotesque to Nuclear-Age Precedents: The Modes and Meanings of Cli-Fi Humour." *Studies in the Novel* 50, no. 1 (2018): 86-107.

Zalewski, Daniel. "The Background Hum: Ian McEwan's Art of Unease." *New Yorker*, 23 February 2009. https://www.newyorker.com/magazine/2009/02/23/the-background-hum.

Zemanek, Evi. "A Dirty Hero's Fight for Clean Energy: Satire, Allegory, and Risk Narrative in Ian McEwan's *Solar*." *Ecozon@: European Journal of Literature, Culture and Environment* 3, no.1 (2012): 51–60. doi: https://doi.org/10.37536/ ECOZONA.2012.3.1.450.

CHAPTER TEN

CARE MORE, DESTROY LESS: THE ENVIRONMENTALIST ETHIC OF CARE IN LIZ JENSEN'S *THE RAPTURE*

IŞIL ŞAHIN GÜLTER

Introduction

Certainly, at the beginning of the twenty-first century, we live in an epoch in which the future, or even the human race's survival, is at stake. Human beings are aware that ecological life on earth has been exploited beyond recognition, and limited restraints have been exercised on that capacity. Human beings' prolonged relationship with nonhuman nature "has largely derived from modernity's irrational fear of nature and hence has created an antagonism between humans and their environments," which Simon Estok refers to as "ecophobia."[1] Estok's definition of 'ecophobia,' emphasizing the irrationality of the antagonism towards nature, demonstrates that the relationship between human and nonhuman nature requires new considerations. As the effects of human conduct on the environment have already been a universal and pervasive condition of the imagination, "the imaginative capacities of the novel have made it a vital site for the articulation of [ecophobia]."[2] The novel has tried to grapple with the realities of the current ecological crisis. By doing so, the novel shows how awareness of this crisis has become part of cultures around the world. In this sense, climate fiction or "literature of *extrapolation*" constructs "the realistic projection of plausible futures from the present" in which environmental problems can not be denied longer.[3] Hence, it should be noted that cli-fi

---

[1] Simon C. Estok, *The Ecophobia Hypothesis* (New York: Routledge, 2018), 1.
[2] Adam Trexler, *Anthropocene Fictions: The Novel in a Time of Climate Change* (Charlottesville: University of Virginia Press, 2015), 23.
[3] Rebecca Evans, "Fantastic Futures? Cli-fi, Climate Justice, and Queer Futurity," *Resilience: A Journal of the Environmental Humanities* 4, no. 2-3 (2017): 99.

provides the crisis's causes, a reconsideration of those causes, and, more importantly, a potential for change.

Liz Jensen's sixth novel, *The Rapture* (2009), set at a near-future moment, employs the techniques of the thriller cli-fi and reflects "apocalyptic global climate change."[4] With an accelerating end-of-the-world scenario, *The Rapture* addresses increasing temperatures, unpredictable weather patterns, frequent hurricanes, tsunamis, earthquakes. Furthermore, *The Rapture* turns attention to the unjust human exploitation of the earth, foregrounding deep-sea drilling as a force of destruction. It can be argued, then, that Liz Jensen participates in Estok's critique of 'ecophobia,' which is central to *The Rapture* and calls for a change that lies in *caring more*. In this context, the chapter first investigates highly contentious gender dimensions of care and then whether care provides a beneficial environmentalist ethic. Liz Jensen's concern for the development of the environmentalist ethic of care[5] can be regarded as vital to creating a sustainable environment since "our construction of 'sustainability' is driven by a notion of care – care for the nonhuman environment enfolded with a concern for our human descendants."[6] Indeed, forging an association between women's caring and environmental politics through maternally and environmentally attuned protagonist Gabrielle Fox, who appears as a *carer* in the book, the main discussion questions whether women's *caring stance* can be a solution to the ecological crisis.

'Care' and 'The Environmentalist Ethic' in the Anthropocene

Anthropocene, popularized by Paul J. Crutzen and Eugene F. Stoermer in early 2000, "emphasize[s] the central role of mankind in geology and ecology" and proposes the fact that the earth has undergone a current geologic phase, which is indeed distinct from the Holocene.[7] In

---

[4] Adam Trexler and Adeline Johns-Putra, "Climate Change in Literature and Literary Criticism," *Wiley Interdisciplinary Reviews: Climate Change* 2, no. 2 (2011): 188; Terry Gifford, "Liz Jensen's *The Rapture* (2009) – Thriller Cli-Fi," in *Cli-Fi: A Companion*, ed. Axel Goodbody and Adeline Johns-Putra (Oxford: Peter Lang, 2018), 150.
[5] The phrase comes from Adeline Johns-Putra, "Care, Gender, and the Climate-Changed Future: Maggie Gee's *The Ice People*," *Green Planets: Ecology and Science Fiction* (2014): 127-142.
[6] Adeline Johns-Putra, "Environmental Care Ethics: Notes Toward a New Materialist Critique," *Symploke* 21, no. 1-2 (2013): 125.
[7] Paul J. Crutzen and Eugene F. Stoermer, "The 'Anthropocene,'" *IGBP Newsletter*

ecocritical circles, the Anthropocene has been received as a cultural concept rather than a geological one. Timothy Morton defines the Anthropocene as "a new phase of history in which nonhumans are no longer excluded or merely decorative features of" human history.[8] In this regard, what the term marks is that human domination over nonhuman nature has reached its apogee. Thus, the agency of the nonhuman actors, including the natural world, must be reconsidered, "bridging the gap of existence – or else they disappear altogether."[9] In a similar vein, Karen Barad mentions that those entities are distinct and independent, which she calls *"agential separability."*[10] To put it more accurately, Barad suggests that the identification of the nonhuman world is not based on a static opposition but on a dynamic process of coming together and enactment of separability, which she refers to as *"agential cut."*[11] Taking Morton, Latour, and Barad's assumptions into consideration, it should be noted that their concern with the ontological questions of agency aims to provide a new framework through which the relationship between human and the nonhuman actors can be reconsidered in the Anthropocene. Adam Trexler uses the term and suggests that "the Anthropocene, by emphasizing a geological process, can usefully indicate the larger, nonhuman aspects of climate."[12] Rather than 'climate change' and 'climate crisis' terms, Trexler prefers to use currently more novel Anthropocene, reflecting the situation that "climate change is upon us."[13] Defining the Anthropocene as a *"cultural transformation"* following the geological transformation of the earth, Trexler pays attention to the immense discourse of climate change and inquiries into how literary studies can be taken as a means of reflecting this phenomenon.[14] Timothy Clark also uses the Anthropocene as a cultural and critical concept and argues that the Anthropocene represents "an irreversible break in consciousness and understanding between the past and present."[15] Treating the Anthropocene

---

41 (2000), 17.

[8] Timothy Morton, *Hyperobjects: Philosophy and Ecology after the End of the World* (Minneapolis: University of Minnesota Press, 2013), 12.

[9] Bruno Latour, "Agency in the Time of the Anthropocene," *New Literary History* 45, no. 1 (2014): 12.

[10] Karen Barad, "Posthumanist Performativity: Toward an Understanding of How Matter Comes to Matter," *Signs: Journal of Women in Culture and Society* 28, no. 3 (2003): 815.

[11] Barad, "Posthumanist Performativity," 815.

[12] Adam Trexler, *Anthropocene Fictions: The Novel in a Time of Climate Change* (Charlottesville: University of Virginia Press, 2015), 4.

[13] Trexler, *Anthropocene Fictions,* 5.

[14] Trexler, *Anthropocene Fictions,* 5.

[15] Timothy Clark, *Ecocriticism on the Edge: The Anthropocene as a Threshold*

as 'a threshold concept', Clark proposes that "the question of [...] the destructiveness of the human species seems [...] a breach of decorum and scale of a kind endemic to the Anthropocene."[16] Alternatively, Astrid Bracke uses the term "climate crisis" rather than the Anthropocene, and she argues that although those terms are not interchangeable, they both encapsulate the problems humanity faces today.[17] It is, therefore, no coincidence that in the Anthropocene, the issues related to the climate have become entrenched in contemporary Western culture.

As the above discussion indicates, humanity's collective hubris lies at the heart of the Anthropocene, and human beings simply do not care for the environment. Many critics have recently re-evaluated the relationships between human and nonhuman nature, arguing that the new considerations are the way forward. In her reconsideration of the relationship between human and the nonhuman world, Dipesh Chakrabarty claims that the main problem which the Anthropocene poses is human beings' unwillingness to understand their roles as geological agents. According to her, without that awareness, "there is no making sense of the current crisis that affects us all."[18] In other words, as Chakrabarty emphasizes, eco-crisis, which emerges as an "unintended consequence of human actions" entails "a new universal history" and "a global approach to politics."[19] In similar terms to Chakrabarty, Haraway suggests, "the Anthropocene marks severe discontinuities; what comes after will not be like what came before. I think our job is to make the Anthropocene as short/thin as possible and to cultivate with each other in every way imaginable epochs to come that can replenish refuge."[20] It is evident that the main challenge of the Anthropocene is that humans can affect, not control. Hence, a new perspective-an environmentalist ethic of care- is required, which encourages the creation of modes of attention that improve mutual understanding of the natural world, promoting appreciation for its existence as a distinct entity. It is, therefore, worthwhile to add one point here: human beings should care more and destroy less, guided by the environmentalist ethic of care, and, more importantly, be allies with nonhuman nature rather than dominators over it.

Concept (London: Bloomsbury, 2015), 130.
[16] Clark, *Ecocriticism on the Edge,* 61.
[17] Astrid Bracke, *Climate Crisis and the 21st Century British Novel* (London: Bloomsbury, 2017), 16.
[18] Dipesh Chakrabarty, "The Climate of History: Four Theses," *Critical Inquiry* 35, no. 2 (2009): 221.
[19] Chakrabarty, "The Climate of History," 222.
[20] Donna Haraway, "Anthropocene, Capitalocene, Plantationocene, Chthulucene: Making Kin," *Environmental Humanities* 6, (2015): 160.

In the relationship between human and nonhuman nature, care ethics has been influential in developing a more hopeful and livable environment for human beings. The notion of the "ethic of care" appeared primarily with Carol Gilligan's *In a Different Voice* (1982) and Nel Noddings's *Caring: A Feminine Approach to Ethics and Moral Education* (1984). The initial theorization of care ethics was based on feminine concerns. Gilligan argues, women conceptualize morality as care rather than fairness and women's moral development is built on "responsibility and relationships" rather than an "understanding of rights and rules."[21] In synch with Gilligan, Noddings developed the notion of care and relationality into distinct components of moral understanding that are "characteristically and essentially feminine."[22] According to Noddings, "an ethic of caring arises out of our experience as women, just as the traditional logical approach to ethical problems arises more obviously from masculine experience."[23] With an emphasis on the latter one, Noddings divides caring activity into two stages, "caring-for" and "caring-about." As Noddings herself puts it, "caring-about is almost certainly the foundation for our sense of justice" and "it moves us from the face to face world into the wider public realm."[24] In her redefinition of care, she calls for an increased responsibility to care about distant humans. More importantly, she defines the natural capacity of caring as "innate to human beings" which can be developed to attain "the ideal of caring" by females and males.[25] Alternatively, Sara Ruddick establishes close connections between care ethics and motherhood, conceiving motherhood as the starting point for the analysis. Ruddick discusses how the practices of maternal people, who may be men or women, demonstrate cognitive capacities or moral conceptions of greater moral significance.[26] Then, it can be argued that the trajectory of multiple directions established the foundations of environmental and explicitly feminist ethics and contributed to the expansion of today's care ethics to a wider scope, including humanity and the environment.

    The ethics of care has been instrumental in building a more hopeful and livable situation for human beings and the nonhuman world. Thus,

---

[21] Carol Gilligan, *In a Different Voice* (Cambridge: Harvard University Press,1982), 19.
[22] Nel Noddings, *Caring: A Feminine Approach to Ethics and Moral Education* (Berkeley: University of California Press, 2003), 29.
[23] Noddings, *Caring,* 29.
[24] Nel Noddings, *Starting at Home: Caring and Social Policy* (Berkeley: University of California Press, 2002), 22.
[25] Noddings, *Caring,* 83.
[26] Sara Ruddick, "Maternal Thinking," *Feminist Studies* 6, no. 2 (1980): 346-348.

critics who focus on the activity of care attach importance to its relationship with environmental concerns. In this sense, Carolyn Merchant emphasizes that women's active role in the global environmental movement today cannot be disregarded. Merchant notes, women may assume the role of bridge-builder between humans and nature, as "women's intimate knowledge of nature" puts forward "a new partnership ethic of earthcare."[27] Forging an association between women's daily caring practices and their concern with the environment, Merchant proposes a socio-material and experiential ecofeminist perspective. In her argument that draws on highly contentious gender dimensions of care and historical connections to the environment, Merchant opposes the homocentric and egocentric ethics of dominant institutions.[28] In synch with Merchant, Chris Cuomo presents one of the most thorough interrogations to date the ecofeminist adoption of the care ethic position. In her argument, she severely opposes biological essentialism, concluding that "asserting that woman = mother, woman = feminine, mother = nature, feminine = caring is not a good idea theoretically and practically."[29] In her opposition, Cuomo raises doubts about the place of feminized notions of care in environmental ethics. Alternatively, Catriona Sandilands argues that the concern women show for the environment is inevitable and simply a product of their location in the private sphere: planetary ecology mirrors housekeeping care.[30] In similar terms to Sandilands, Val Plumwood labels this typology "the angel in the ecosystem."[31] Sandilands proposes, coining the term "motherhood environmentalism," that highly controversial gender dimensions of care should be re-evaluated before being adopted as a model for political action.[32] To put it more clearly, Sandilands suggests that instead of clinging to 'women' as subjects with inherent qualities that are pre-political, the ecofeminists might take a perspective that acknowledges the multiplicity of subject positions.[33] In Sandilands's words:

---

[27] Carolyn Merchant, *Earthcare: Women and the Environment* (New York: Routledge, 1995), 16, 209.
[28] Merchant, *Earthcare,* 215-216.
[29] Chris J. Cuomo, *Feminism and Ecological Communities: An Ethic of Flourishing* (New York: Routledge, 1998), 126.
[30] Catriona Sandilands, *The Good-Natured Feminist: Ecofeminism and the Quest for Democracy* (Minneapolis: University of Minnesota, 1999), xii.
[31] Val Plumwood, *Feminism and the Mastery of Nature* (New York: Routledge, 1993), 9.
[32] Sandilands, *The Good-Natured Feminist,* xiii.
[33] Sherilyn Macgregor, *Beyond Mothering Earth: Ecological Citizenship and the Politics of Care* (Vancouver: UBC, 2006), 53.

> What the idea of speech in identity politics highlights is a form of democratic subjectivation – the construction of political consciousness, an 'I,' through the necessary uncovering and reconstruction of ways of being in the world oppressed and marginalized by hegemonic political and cultural formations.[34]

As the quotation indicates, for a woman, the ability to speak for herself rather than be spoken for is a means of finding a new voice in the privileged male world of speech. Thus, as Sandilands mentions, the process of 'subjectivation' enables women to act politically and appear as citizens in public that will allow them to express their identity "that cannot be grounded in or predicted by private life."[35] In other words, Sandilands favors women's politicization through environmental struggles as environmentalism is possible via 'subjectivation' and our care reaction to any nonhuman phenomenon runs parallel to this subjectivating urge.[36]

Particularly, Sherilyn Macgregor develops Sandilands's identification of women's politicization through environmental struggles. Coining the term "ecomaternalism,"[37] she attempts to theorize women's engagement in ecopolitics and advocates Sandilands's 'democratic subjectivation,' which provides women with political citizenship. With a clear emphasis on women's caring stance, Macgregor questions whether care – as in earthcare- can be held up as a solution to the environmental problems.[38] As Merchant, Cuomo, Sandilands, and Macgregor mention, the relationship between women's daily caring practices and their concern with the environment needs to be reconsidered in a new theoretical framework. Instead of clinging to women as subjects with inherent qualities that are pre-political, the new framework would expand the private and direct mode of care into a rationale for political and indirect care for the future. To put it differently, women's caring stance towards the wider social and biophysical worlds would undoubtedly become part of a caring infrastructure that helps future generations. In the light of those preliminary observations of care and environmental ethics in the Anthropocene, the following discussion will revolve around the protagonist Gabrielle Fox's attempts to develop a caring stance both in the private/maternal and public/environmental sphere in *The Rapture*.

---

[34] Sandilands, *The Good-Natured Feminist*, 80.
[35] Sandilands, *The Good-Natured Feminist*, 160.
[36] Adeline Johns-Putra, "Environmental Care Ethics: Notes Toward a New Materialist Critique," *Symploke* 21, no. 1-2 (2013): 130.
[37] Macgregor, *Beyond Mothering Earth*, 3.
[38] Macgregor, *Beyond Mothering Earth*, 58.

The Environmentalist Ethic of Care in *The Rapture*

The Rapture is set in 2013 England and depicts a version of life in the style of a dystopian novel. From the first pages of the novel, extreme temperature rises of up to "forty in the shade," "maverick weather" events, and drought establish the parameters of climate change.[39] In this sense, Jensen's depiction can serve as a springboard for thinking about recent climate crisis writings engaged with both personal and global paradigms of crisis. On a personal level, as the narrator's inner voice proposes, the book presents no sense of the future and uncertainty of the present. On a global level, Jensen depicts a world that is experiencing "'interesting times,' [with] its food shortages and mass riots and apocalyptically expanded Middle East war, [...] the Faith Wave that followed the global economic crash."[40] Moreover, the world is running out of oil, a circumstance, which would precipitate the final crisis. Jensen's portrayal of a planet that is no longer changing but is running out of natural resources could be interpreted as a Malthusian warning that even the most egalitarian societies would inevitably return to conflict and competition for scarce resources.[41] In this context, amidst the ecological disaster and evangelical millenarism, *The Rapture* develops two different readings for the end of the age from the perspectives of the faith-wavers and the ecologists navigating between twin horizons of an apocalypse. On the one hand, the faith wavers expect the Rapture after which the novel is entitled, "a notion that is heavily debated in the Faith Wave's Armageddon discussions, along with the belief that the Messiah will return after a seven-year period of 'Tribulation' or 'End Times', in which God will punish humankind for its sins, by means of plagues, floods, fire, and brimstone, etc."[42] On the other hand, following the failure in Copenhagen climate delivery, an eco-group called Planetarians began to preach their own cataclysmic vision of human extinction and welcome the many climatic catastrophes as a form of "human cull."[43] According to Haris Modak, the spiritual leader of the Planetarian movement, whose assumptions are based on James Lovelock's *Gaia theory*, "the planet [is] a self-regulating organism with its own 'geophysiology.'"[44] Gabrielle Fox, the narrator, adheres to neither of those ideologies. However, she

---

[39] Liz Jensen, *The Rapture* (London: Bloomsbury, 2009), 3-4.
[40] Jensen, *The Rapture,* 10.
[41] Greg Garrard, *Ecocriticism,* 2nd ed. (London and New York: Routledge, 2012), 102.
[42] Jensen, *The Rapture,* 35.
[43] Jensen, *The Rapture,* 36.
[44] Jensen, *The Rapture,* 36.

encounters the belief systems of both faith wavers and planetarians on different occasions. Both the faith-wavers who have embraced climate change as "a sign we're on the brink of doomsday"[45] and the Planetarians offer their own apocalyptic vision, and they believe that humans are living through the end of days; thus the Rapture is near.

Gabrielle has been appointed to Oxsmith Adolescent Secure Psychiatric Hospital to carry out Bethany Krall's treatment. Bethany, daughter of an evangelical preacher father, "during the Easter school holidays [...] stabbed her mother Karen to death in a frenzied and unexplained attack."[46] Bethany was diagnosed with Cotard's syndrome, in which the patient believes s/he is already dead, and she gave no response to any of her therapy programs. Bethany systematically rejects treatment which will eventually lead to her being treated by electroconvulsive therapy that "gave her relief from the delusion of being dead but stimulated a preoccupation with climate change, chemical pollution, weather patterns, geological disturbances, and apocalyptic scenarios."[47] Electroconvulsive therapy reveals that Bethany can reliably predict the dates of natural disasters. Assigned to Bethany, who has "satellite vision. Like the Hubble telescope", as her art therapist, Gabrielle is at first sceptical about Bethany's "[unimaginative] cataclysmic visions"[48] since recent TV shows, documentaries, movies all refer to worldwide ecological disasters, and those visions have already been shared by half of the population. Gabrielle attempts to evaluate Bethany's apocalyptic visions by analyzing her unconscious, "get[ting] to the bottom of her."[49] Bethany is resistant and warns Gabrielle that "This is what you don't get [...] It's about what's going to happen."[50] Gabrielle employs different strategies, including art therapy, listening techniques, silence, all of which are ignored by Bethany, who remains "like a nagging crossword clue that [she] can't solve."[51] However, Gabrielle and Bethany develop a mother-child relationship as Bethany allows Gabrielle to "pretend to be [her] mum."[52] It is, therefore, no coincidence that Gabrielle engenders a caring stance towards Bethany and responds to her needs, as the "ethics of care stresses the moral force of the responsibility to respond to the needs of

---

[45] Jensen, *The Rapture*, 88.
[46] Jensen, *The Rapture*, 9.
[47] Jensen, *The Rapture*, 35.
[48] Jensen, *The Rapture*, 25.
[49] Jensen, *The Rapture*, 49.
[50] Jensen, *The Rapture*, 42.
[51] Jensen, *The Rapture*, 49.
[52] Jensen, *The Rapture*, 58.

the dependent."[53]

By the way, the flashbacks inform the reader about Gabrielle's paraplegic recovery from a traumatic car accident, the death of her lover, the loss of her unborn child, and her inability, with her disability, to have any more children. She describes herself as "a non-woman pretending to be a real one [and] a woman with no man, no baby, no feeling below the waist, no imaginable future"[54] to which Johns-Putra refers to as "a self-description which collapses various conventional idealisations of female identity into one, all imbricated in the physical (motherhood, (hetero) sexual desirability and desire) and all deemed necessary to a life worth living (an 'imaginable future')."[55] However, childlessness represents the most tragic part of Gabrielle's deteriorated sense of self, which comes to the surface in her relationship with Bethany. Gabrielle's caring stance towards Bethany, drawing on highly contentious gender dimensions of care, displays her maternal desire, which constructed Bethany's treatment "in affectionate and even caring terms."[56] Gabrielle's initial attitude towards Bethany is based on professional concern as she states that "a part of me that's still professional cares."[57] She wants Bethany to be punished with isolation due to her assault of another patient, during which Gabrielle herself had a head injury. Then, Gabrielle's professional care extends to maternal care when Bethany tells her about the abuse that drove her to murder her mother. That is, Bethany's exposure to physical and emotional abuse from her parents creates a space for Gabrielle's emotional attachment to Bethany. In this sense, while Gabrielle functions as a mother figure that cares about Bethany's feelings, Leonard Krall and Karen Krall are depicted as abusive parents. Bethany's detailed recounting of physical and emotional abuse that drove her to murder her mother reveals how her parents are insufficient to care for Bethany. As Held points out, "affectionate sensitivity and responsiveness to the need" are the vital requirements of care ethics, which would provide better moral guidance.[58] As it can be inferred from Bethany's moral deterioration and mental instability showing up in her threats and violence against others, including Gabrielle, Bethany's parents could not meet her needs and take responsibility for her. Held mentions, "the central

---

[53] Virginia Held, *The Ethics of Care: Personal, Political, and Global* (Oxford: Oxford University Press, 2006), 10.
[54] Jensen, *The Rapture,* 65, 109.
[55] Adeline Johns-Putra, *Climate Change and the Contemporary Novel* (Cambridge: Cambridge University Press, 2019), 102.
[56] Johns-Putra, *Climate Change,* 102.
[57] Jensen, *The Rapture,* 103.
[58] Held, *The Ethics of Care,* 24.

focus of the ethics of care is on the compelling moral salience of attending to and meeting the needs of the particular others for whom we take responsibility."[59] Taking care of one's child, for instance, should be at the top of a person's moral interests in a good and reasonable way. Bethany's case creates a space for Gabrielle, who accepts her with sympathy and finds herself drawn to her. She refers to maternal responsibility and tells Bethany that "your mother's job was to protect you. That's what parents are supposed to do. What they did to you was wrong."[60] She criticizes maternal neglect and assumes maternal protection for Bethany as her inner narrative advises, "If Karen Krall were standing in front of me now, perhaps I'd want to kill her myself."[61] During her institutionalization at Oxsmith, "Bethany Krall had made four attempts on her own life […] She began to starve herself […] Bethany showed no interest in co-operating in therapy sessions, and remained largely mute."[62] Upon Gabrielle's maternal sensitivity and responsiveness to her feelings, Bethany confides in her, and they develop a relationship that would contribute to the progress of human beings. As Gilligan argues, "in the different voice of women lies the truth of an ethic of care, the tie between relationship and responsibility,"[63] which would construct a healthier and more harmonious world. In this respect, Gilligan's assumptions sowed the seeds of care ethics and, arguably, earthcare ethics, allowing women to enter the public domain. In synch with Gilligan, Noddings suggests, natural caring sentiment in a mother-child relationship is followed by ethical caring "in response to a remembrance of the first."[64] Thus, it should be noted that Gabrielle's caring practice as a surrogate mother in her direct private relationship with Bethany has been expanded to an indirect broader scope as a citizen of the world in the course of the novel. Considering this point of view, Gabrielle's maternal care for Bethany should therefore become an instrument of environmental care with an emphasis on women's participation in the environmental struggle. Caring relationships that reach beyond the narrower personal circumstances can be extended to the broader global arena, transforming a caring person into a caring citizen of the world.

The electroconvulsive therapy stimulates Bethany's eco-apocalyptic visions, which are significant to the discussion on Gabrielle's participation in environmental struggle. As mentioned above, during electroconvulsive

---

[59] Held, *The Ethics of Care*, 10.
[60] Jensen, *The Rapture*, 275.
[61] Jensen, *The Rapture*, 275.
[62] Jensen, *The Rapture*, 11.
[63] Gilligan, *In a Different Voice*, 173.
[64] Noddings, *Caring*, 79.

therapy sessions, Bethany can foretell dates and locations of certain ecological disasters without knowing the specifics. She describes the perils of nature which have claimed thousands of lives. Gabrielle establishes a relationship between Bethany's bodily experiences during electroconvulsive therapy and the damage inflicted on the earth, deducing that: "Bethany's pain is planet-shaped and planet-sized: she has her own vividly imagined earthquakes and hurricanes, her own volcanic eruptions, her own form of meltdown."[65] In other words, as Jensen's eco-thriller progresses, Bethany shows herself to be a passive medium for the planet's misery, with her body acting as a microcosm and the earth serving as a macrocosm. It appears that her electroconvulsive therapy sessions allow her to connect with the earth and feel what the earth is going through inside her body.[66] Bethany predicts a hurricane in Brazil on 29th July, an earthquake in Istanbul, and a tornado in the American Midwest. Then, she adds that "there is going to be a tornado in Scotland any day now […] And the big one's on its way. The Tribulation starts in October."[67] In fact, Bethany correctly predicts the effects of those natural cataclysms, despite her ignorance of the effects. Thus, it can be noted that the careless and insensitive relationship between Bethany and her parents induced her "moral failure of ignorance –or, worse, indifference" to her environment.[68] In Held's terms, she fails to build "care and concern and mutual responsiveness on both the personal and wider social levels."[69] Bethany goes on to comment that "I want them to die. The planet's overpopulated, right? […] *The fewer the merrier. More oxygen for the rest of us. Organic diseases.*"[70] Bethany unconsciously adopts the zero-birth rate discourse, which is proposed by the Planetarians as a solution to the overpopulation problem. Haris Modak writes as follows in one of his articles:

> We are the agents of our own destruction – and when we are gone, extinguished by our own heedless quest for expansion, the planet will not mourn us […] Today, the human species stands at the brink of a new mass extinction […] it would seem that the kindest thing to do for our grandchildren is to refrain from generating them.[71]

---

[65] Jensen, *The Rapture,* 36.
[66] Helen E. Mundler, *The Otherworlds of Liz Jensen: A Critical Reading* (Rochester, New York: Camden House, 2016), 164.
[67] Jensen, *The Rapture,* 48-49.
[68] Ornaith O'Dowd, "Care and Abstract Principles," *Hypatia* 27, no. 2 (2012): 419.
[69] Held, *The Ethics of Care,* 43.
[70] Jensen, *The Rapture,* 43.
[71] Jensen, *The Rapture,* 37.

According to the Planetarians, overpopulation is one of the most critical aspects of the ecological breakdown, and epidemics are the natural outcomes of Gaia's self-regulated system to combat population growth.[72] Indeed, the Planetarians believe that combating organic illnesses encourages population growth, which exacerbates ecological problems. As mentioned above, Bethany participates in the zero-birth rate argument and clearly illustrates that there is no point in having a baby in such a world in which Bethany depicts even her imaginary country, Bethanyland, as a "suck[ing] ... and completely fucked-up place [where] the trees are all burned. Everything is poisonous."[73] In this way, the novel establishes the issue of overpopulation as a contributing factor to the environmental crisis. The Planetarians' zero-birth rate discourse is also supported by Gabrielle, who reconsiders "people's right to have children" and develops a defence against overpopulation.[74] Johns-Putra notes that Gabrielle's profound maternal instincts complicate the novel's eco-centric position on posterity, with its dreams of a nearly childless and probably nonhuman future. However, it is significant to note that Gabrielle develops a caring stance towards the environment, which would secure an alternative future for humans and nonhumans as the novel progresses.[75]

As the disasters in question occur on the expected days, Gabrielle feels guilty for not alerting the world's population about the potential ecological disasters. Therefore, Gabrielle seeks physicist Fraser Melville's assistance, who provides a scientific basis for Bethany's susceptibility to turbulence structures and the natural activities in the air, the seas, and the earth itself. Fraser suggests that Bethany experiences the natural activities on the earth within her body during electroconvulsive therapy, and he deduces, "Bethany's some kind of New Age eco-psychic."[76] Fraser and Gabrielle analyze Bethany's drawings which refer to a set of dates, places, and events, and then they realize that she did predict all those events accurately. The next event, "Tribulation" is scheduled for 12th October and involves the exploration of the North Sea seabed for frozen methane.[77] The methane crisis scenario, which is brought to light near the end of the novel by Bethany, is expertly summarized by ecologist Ned as follows:

---

[72] Jensen, *The Rapture*, 33.
[73] Jensen, *The Rapture*, 47.
[74] Jensen, *The Rapture*, 34.
[75] Johns-Putra, *Climate Change*, 101-102.
[76] Jensen, *The Rapture*, 136.
[77] Jensen, *The Rapture*, 135.

> Since the energy companies started trying to exploit the sub-oceanic hydrates, the drilling's increased the threat. Dramatically. Post-peak oil, everyone's after it. China, the US, India. Hundreds of experimental rigs, planted off coastlines all round the world.[78]

Then, he warns that the seabed extraction to supply potential oil reserves poses many risks, including the release of massive quantities of explosive methane and the destabilization of the seabed itself, resulting in a series of underwater avalanches. Ned further speculates that the scenario may be followed by "runaway global warming on a scale that's beyond anyone's worst nightmare."[79] The third part of the novel revolves around Bethany's methane disaster prophecy and Gabrielle's efforts to co-operate with the international scientific authorities and eco-groups to prevent this particular disaster.

More notably, Gabrielle's concern for Bethany is linked to her concern for the environment. Gabrielle not only wants to keep Bethany safe and alive, but she also wants to save as many human lives as she can. Her initial attempt is to feel responsible for the distant others, conforming to Garrard's conclusion that "only if we imagine that the planet has a future, after all, are we likely to take responsibility for it."[80] Even though Terry Gifford argues that while the world promises a future at the end of the novel, no one seems to be able to take responsibility for it at any stage in the novel,[81] my reading suggests that Gabrielle is aware that "for progress to be made, persons need to care together for the well-being of their members and their environment."[82] It is, therefore, no coincidence that Gabrielle takes on the responsibility regarding environmental care and endeavours to recruit the Planetarians to her cause. She debates the ethical value of warning the public about impending threats with Haris Modak, who "truly believes the world will be a better place without humans, and sees time in terms of epochs rather than days and hours, then yes: why should he bother to save a few random millions."[83] In this respect, Gabrielle is "positioned as a potential savior of the planet's human beings"[84] when she criticizes Modak's indifference to this immanent threat and rebukes him as follows:

---

[78] Jensen, *The Rapture*, 212.
[79] Jensen, *The Rapture*, 228.
[80] Garrard, *Ecocriticism*, 107.
[81] Terry Gifford, "Biosemiology and Globalism in *The Rapture* by Liz Jensen," *English Studies* 91, no. 7 (2010): 726.
[82] Held, *The Ethics of Care*, 43.
[83] Jensen, *The Rapture*, 249.
[84] Johns-Putra, *Climate Change*, 103.

> The issue is about the people who are alive now, who will die if you don't help us warn them! [...] If we fail to act now, none of us is any better than any criminal on trial in The Hague. Most of all you, because you're the one with the power to do something.[85]

Indeed, this responsibility is to be celebrated because caring for distant others and environments produces unique insights into the interrelated processes of life. Here, Gabrielle mediates the connection between humans and the environment. She does this because she believes that care should be seen as integral to any notion of a good society. On the one hand, she applies maternal and caring values to environmental problems as a form of 'earthcare' as Merchant identifies. On the other hand, she privileges women's experiential knowledge as a solution to the ecological crisis by arguing for a public space wherein women as citizens can raise and demand deliberation on its merits. In Sandilands's words, Gabrielle launches "a new series of codes through which to perceive and act in the world and through which to challenge and therefore change dominant and oppressive constructions of sense."[86]

Bethany's prophetic vision of the methane disaster sets the scene for a climax filled with apocalyptic spectacle. The date was also regarded as the day of the Rapture by the faith wavers who gathered at the post-2012 Olympic stadium in East London for the Rapture event. On the verge of methane disaster, as the earthquake and tsunami threaten, Bethany, Gabrielle, and Frazer head to the Olympic stadium, where the other scientists have arranged for a helicopter to fly them out of the upcoming flood. The three are rescued just in time, being lifted out of the stadium minutes before "the fire spreads greedily as though devouring pure oil, yellow flames bursting from the crest of the liquid swell, triggering star-burst gas explosions above [...] The heat is unbearable [...] It is almost impossible to breathe."[87] *The Rapture* envisions an apocalyptic ending with Bethany's suicide, falling from the helicopter, and Gabrielle's hopeless attitude towards life as she articulates:

> I know already there will be no green fields in Bethanyland, no safe place for a child to play. Nothing but hard burnt rock and blasted earth, a struggle for water, for food, for hope. A place where every day will be marked by the rude, clobbering battle for survival and the permanent endurance of regret, among the ruins of all we have created and invented, the busted remains of the marvels and

---

[85] Jensen, *The Rapture*, 256.
[86] Sandilands, *The Good-Natured Feminist*, 103.
[87] Jensen, *The Rapture*, 339.

> commonplaces we have dreamed and built, strived for and held dear: food, shelter, myth, beauty, art, knowledge, material comfort, stories, gods, music, ideas, ideals, shelter […] I look out on to the birthday of a new world. A world a child must enter.
> A world I want no part of.
> A world not ours (341).

As the quotation indicates, *The Rapture* presents a stark and uncomfortable future referring to the human destructiveness in the Anthropocene. However, the point that needs to be foregrounded is that the apocalyptic spectacle aims at the assumption that the destruction of the earth can be avoided. As Garrard points out, "apocalyptic rhetoric […] is capable of galvanising activists, converting the undecided and ultimately, perhaps, of influencing government and commercial policy."[88] Jensen's bleak depiction of the radical deterioration of the environment, resulting from risky technologies, emphasizes the threats to both humanity and the environment, which would eventually head towards a deadly climax that obliterates both. Thus, human beings would opt for sustainable technology that would let them save the ecologically degraded planet. In this sense, *The Rapture* invites the readers to the awareness that "Humans, who have the power to destroy nonhuman nature and potentially themselves through science and technology, must exercise care and restraint by allowing nature's beings the freedom to continue to exist."[89] As Merchant reinforces, environmentalist care ethics should be created in which humans pay equal attention both to basic human needs and to the voice of nature by restraining humanity's arrogance.

Conclusion

The Anthropocene marks that human domination over nonhuman nature is now complete, and the relationships between those entities require new considerations. As the above discussion has indicated, a new framework through which the agency of nonhuman actors can be reconsidered and the appreciation of it as a distinct entity can be encouraged, is the way forward. The notion of care promises great hope for the reconsideration of human-nonhuman relationships and the creation of a more sustainable planet. In this regard, much importance has been attached to women's caring stance, which may be positioned as a solution to the ecological crisis. The terms "earthcare," "motherhood environmentalism,"

---

[88] Garrard, *Ecocriticism,* 113.
[89] Merchant, *Earthcare,* xix.

"angel in the ecosystem," and "ecomaternalism," have been coined not only to oppose clinging women to the essentialist qualities that are pre-political but also to theorize the feminized notions of care through a more political frame. In this way, the vitality of women's engagement in ecopolitics, finding a new voice in the privileged male world of speech, has been foregrounded. Thus, a celebratory link between women's environmental activism and their feminine or maternal instincts has been brought forward. As the above discussion has shown, women's caring stance has been suggested as a solution to the ecological crisis in the Anthropocene. The tendency to invoke the environmentalist ethic of care as grounds forges the assumption that the environment might be saved with special emphasis on feminine care as a viable alternative. More importantly, the fact that Gabrielle has been empowered through her care-inspired eco-activism may be accompanied by a sufficient consideration of environmental transformation. *The Rapture*, foregrounding deep-sea drilling as a force of destruction, offers an entry point to transform the environment through a partnership between human and nonhuman nature against the destructiveness of human beings. In this way, *The Rapture* can engender a deeper influence on the awareness of environmentalist care ethics. As a thriller cli-fi, *The Rapture*, calls for a transition from 'ecophobia' to 'earthcare,' led by Gabrielle, whose caring stance extends from the private maternal sphere to the global environmental one. Considering this point of view, the apocalyptic spectacle, which *The Rapture* envisions, should be regarded as a rallying call to reconfigure humans' attitudes in the present in order to guarantee the well-being of humans and nonhumans in the future.

References

Barad, Karen. "Posthumanist Performativity: Toward an Understanding of How Matter Comes to Matter." *Signs: Journal of Women in Culture and Society* 28, no. 3 (2003): 801-831.

Bracke, Astrid. *Climate Crisis and the 21st Century British Novel.* London: Bloomsbury, 2017.

Chakrabarty, Dipesh. "The Climate of History: Four Theses." *Critical Inquiry* 35, no. 2 (2009): 197-222.

Clark, Timothy. *Ecocriticism on the Edge: The Anthropocene as a Threshold Concept.* London: Bloomsbury, 2015.

Crutzen, Paul J. and Eugene F. Stoermer. "The 'Anthropocene.'" *IGBP Newsletter* 41, (2000): 17-18.

Cuomo, Chris J. *Feminism and Ecological Communities: An Ethic of Flourishing.* New York: Routledge, 1998.

Estok, Simon C. *The Ecophobia Hypothesis*. New York: Routledge, 2018.
Evans, Rebecca. "Fantastic Futures? Cli-fi, Climate Justice, and Queer Futurity." *Resilience: A Journal of the Environmental Humanities* 4, no.2-3 (2017): 94-110.
Garrard, Greg. *Ecocriticism.* 2nd Ed. London and New York: Routledge, 2012.
Gifford, Terry. "Biosemiology and Globalism in *The Rapture* by Liz Jensen." *English Studies* 91, no. 7 (2010): 713-727.
Gifford, Terry. "Liz Jensen's *The Rapture* (2009) – Thriller Cli-Fi." In *Cli-Fi: A Companion*, edited by Axel Goodbody and Adeline Johns-Putra, 147-152. Oxford: Peter Lang, 2018.
Gilligan Carol. *In a Different Voice*. Cambridge: Harvard University, 1982.
Haraway, Donna. "Anthropocene, Capitalocene, Plantationocene, Chthulucene: Making Kin." *Environmental Humanities* 6, (2015): 159-165.
Held, Virginia. *The Ethics of Care: Personal, Political, and Global*. Oxford: Oxford University Press, 2006.
Jensen, Liz. *The Rapture*. London: Bloomsbury, 2009.
Johns-Putra, Adeline. "Environmental Care Ethics: Notes Toward a New Materialist Critique." *Symploke* 21, no. 1-2 (2013): 125-135.
Johns-Putra, Adeline. *Climate Change and the Contemporary Novel*. Cambridge: Cambridge University Press, 2019.
Latour, Bruno. "Agency in the Time of the Anthropocene." *New Literary History* 45, no. 1 (2014): 1-18.
Macgregor, Sherilyn. *Beyond Mothering Earth: Ecological Citizenship and the Politics of Care*. Vancouver: UBC, 2006.
Merchant, Carolyn. *Earthcare: Women and the Environment*. New York: Routledge, 1995.
Morton, Timothy. *Hyperobjects: Philosophy and Ecology after the End of the World*. Minneapolis: University of Minnesota, 2013.
Mundler, Helen E. *The Otherworlds of Liz Jensen: A Critical Reading*. Rochester, New York: Camden House, 2016.
Noddings, Nel. *Starting at Home: Caring and Social Policy*. Berkeley: University of California Press, 2002.
Noddings, Nel. *Caring: A Feminine Approach to Ethics and Moral Education*. Berkeley: University of California Press, 2003.
O'Dowd, Ornaith. "Care and Abstract Principles." *Hypatia* 27, no. 2 (2012): 407-422.
Plumwood, Val. *Feminism and the Mastery of Nature*. New York: Routledge, 1993.
Ruddick, Sara. "Maternal Thinking." *Feminist Studies* 6, no. 2 (1980): 342-367.

Sandilands, Catriona. *The Good-Natured Feminist: Ecofeminism and the Quest for Democracy*. Minneapolis: University of Minnesota, 1999.

Trexler, Adam and Adeline Johns-Putra. "Climate Change in Literature and Literary Criticism." *Wiley Interdisciplinary Reviews: Climate Change* 2, no. 2 (2011): 185-200.

Trexler, Adam. *Anthropocene Fictions: The Novel in a Time of Climate Change*. Charlottesville: University of Virginia Press, 2015.

Chapter Eleven

Monstrous Agency and Apocalyptic Visions in Paulo Bacigalupi's Climate Change Trilogy: *Ship Breaker*, *Drowned Cities*, and *Tool of War*

Sukanya B. Senapati

Introduction

When eminent scientists drew attention to escalating global temperatures and warned about future Earth disasters, creative writers joined forces to draw attention to the urgency of the crisis by developing a new genre of literature devoted to climate change: climate fiction (cli-fi). Since the 1980s, cli-fi writers have entangled with the science of climate phenomena and imagined Earth scenarios altered by climate change to warn Homo sapiens of the consequences of callous and unsustainable resource gluttony. However, while advances in science and technology are opening up extensive vistas and intricacies of climate science, the ubiquitous use of technology that is rapidly overtaking all aspects of human life, is, also, causing a sense of zombie paralysis that is undermining our species' sense of perceived agency in mitigating anthropogenic climate change. Nonetheless, both climate scientists and avant-garde cli-fi writers are making heroic attempts to engage all people in the discourse of climate change, first, by consciously practicing the non-separation of nature and culture and, secondly, by debunking the myth of human exceptionalism, thereby, compelling Homo sapiens to view themselves as one of innumerable life forms on Earth. As Donna Haraway proposes, entanglements in multi-modalities are necessary to develop sustainable relationships with Earth, as a living, breathing web of incredible complexity that James Havelock termed Gaia. Dualistic, hierarchical thinking, that arbitrarily, and, sometimes, politically values one term over the other in the nature/culture dyad, has been identified as one of many kinds of cultural practices that led to notion of Earth's inexhaustibility, to be used and abused indefinitely, without any adverse consequences. Additionally, because the human has traditionally been separated from and valued over the non-human, Homo sapiens' inescapable dependency on the non-human for sheer existence has been rendered invisible and thus misperceived as unnecessary.

While cli-fi writers are not the first to draw attention to the power of 'Nature,' they are, perhaps, pioneers in their attempt at reimaging Earth, not as a backdrop for human culture, but human culture as always and already imbricated in Earth's materiality. For example, while Shakespeare in *King Lear* (1606) demands that the ruler, King Lear, face his own negligence in protecting his subjects, especially his poor, disenfranchised subjects, like poor Tom, from the vicissitudes of erratic weather, Paulo Bacigalupi imagines Earth futures where the majority of humans, except the privileged and powerful few, as battered and burned by unceasing inclement weather. Such writers' goal is to compel humans to confront the consequences of their own suspension of intelligence and will, as Earth recalibrates itself to reach a different state of equilibrium that will most likely not be conducive to the kind of human life Homo sapiens have become accustomed to in a very short span of geological time. Additionally, long before cli-fi writers made their appearance on the literary scene in the last few decades of the twentieth century, the Romantic Movement had drawn attention to Earth's environment, termed 'Nature' during that time, as necessary to human emotional and spiritual wellbeing. Samuel Taylor Coleridge, one of the founding members of the Romantic Movement, was, also, perhaps, one of the first literary writers to draw attention to the callous abuse and slaughter of animals that has, also, contributed to Earth devastations. In *The Rime of the Ancient Mariner* (1834), Coleridge presents the Sisyphus-like activity of an Ancient Mariner, who is compelled to confess, over and over again, his callous slaughter of an albatross, for no reason at all, *but just because he could*. For his egregious crime, he is punished through the death of all his fellow sailors, and remains stranded, alone at sea, with no fellowship with anybody or anything. Reprieve comes to him only when he learns to love creatures that humans have generally regarded as beneath their acknowledgement, or viewed as ugly and creepy, such as slithering snakes and slimy sea creatures, for that is *all* of the living creatures that are left in the world that he inhabits.

More recently, in Yann Martel's *Life of Pi* (2001), a significant and popular novel of the twentieth-first century, animals, especially a Bengal tiger, take center stage. The protagonist, Pi, is taught two unusual lessons about animals by his zookeeper father: animals are dangerous and Homo sapiens are the most dangerous of them all. Pi's father educates Pi by making his family watch as tigers are being fed, lest familiarity with ferocious animals interned at zoos cause amnesia about animal difference and power; the second that Homo sapiens are the most dangerous animals is meant for all visitors to the zoo for on the wall behind the ticket counter, in bright, red, all capital letters, visitors are asked the question: "Do you know which is the most dangerous animal in the zoo?"[1] An arrow points to the answer that lies concealed behind a curtain which when drawn, reveals the visitor's own visage on a mirror. This warning about humans, specifically one's own self as the most dangerous animal begs the question: dangerous to whom or what? The answer as cli-fi texts expose is that humans are a threat to all of creation, including and especially to their own collective, narcissistic selves.

Contemporary novelist, Paulo Bacigalupi, in his cli-fi trilogy consisting of the *Ship Breaker* (2010), *The Drowned Cities* (2012), and *Tool of War* (2017), explores the horrifying consequences of unfettered objectification of Earth materials and life forms alike, along with notions of ownership that accompany such objectifications and lead to Earth holocaust. Additionally, Bacigalupi interrogates the unexamined use and abuse of science and technology that not only de-calibrate optimal Earth system settings, but experiment with basic building blocks of life. The unanticipated and dangerous consequences of gene editing, even if they are made for benevolent purposes - arguments often used to justify gene therapy - are presented in the trilogy as an opened, Pandora's box of unprecedented chaos. Homo sapiens' objectification of the material and the living alike, along with their indiscriminate use and abuse of science and technology that exacerbate climate change, are all presented in the trilogy as occurrences that have already happened; thereby, forcing readers to pause and contemplate the fossil-fueled, frenzied activities of contemporary, modern living that bequeaths a horrendous environment to both current and future Homo sapiens. Consequently, the so-called intelligent species is exposed as the non-intelligent and destructive one, incapable of acting with purpose, wisdom, and moral rectitude, in spite of all of its accumulated knowledge and advances in science and technology.

---

[1] Yann Martel, *Life of Pi*. (Orlando; Austin; New York: Harvest-Hardcourt, 2001), 34.

Paulo Bacigalupi's Climate Change Trilogy

While Rob Nixon in his non-fictional book, *Slow Violence and the Environmentalism of the Poor* (2011), exposes the detailed reality of the disenfranchised poor who disproportionately experience the worst effects of climate disasters, Bacigalupi in his fictional trilogy imagines future Earth scenarios that parallel and compound climate change consequences and the disenfranchised poor's horrors, thereby, adding an emotional and cultural dimension to the climate change phenomena. Bacigalupi's trilogy features the production of a limited number of powerful creatures, variously called "half-men," "augments," "tools," "killer dogs," etc., that are ultimate killing machines created from a "genetic cocktail of humanity, tigers, and dogs."[2] Had the enormous strength and power of these half-men been engineered to substitute for child or animal labor, their existence could, possibly, have been justified; however, in the trilogy, the disenfranchised poor and the starving children toil in increasingly larger numbers, in impossibly dangerous working conditions, because in the climate-devastated drowned cities, abandoned and orphaned children, called "licebiters,"[3] are routinely tortured and recruited by competing war lords for illicit and dangerous work that pays little to no wages.

The climate devastated world in Bacigalupi's trilogy is set in a not-too-distant future, in the post Accelerated Age, when "much of the world has ... collapsed under disasters. Droughts and floods. Hurricanes, Epidemics and crop failures ... Starvation and refuge wars"[4] are the new normal. During this time, nation states no longer exist and the world is run by multi-national corporations that are protected by international security companies whose modus operandi is violence. What was once the United States is now roughly divided into the North and South with the rich and the privileged known as "swanks"[5] constantly moving up North onto higher ground as more and more coastal cities and low lying areas are drowned by rising seas. In such inhospitable environments, most of the human population has "settled into poverty,"[6] and eke out whatever existence they can in the appropriately named Drowned Cities of the South. Paradoxically, in this post Accelerated Age there are "enough drowned cities and enough people dying from drought," and category six storms, called "city killers," routinely devastate coastal cities on river deltas as the poor just wait for the

---

[2] Paulo Bacigalupi, *Ship Breaker* (New York, Boston: Little Brown, 2010), 211-12.
[3] Ibid, 46.
[4] Paulo Bacigalupi, *Tool of War* (New York, Boston: Little Brown, 2017), 35.
[5] Paulo Bacigalupi, *Ship Breaker*, 287.
[6] Ibid, 217.

worst of "nature's violence to pass."[7] In the aftermath of its passing, the debris of an Earth holocaust is left behind for routine clean-up. In such a world, wealth and power reside in the hands of "swanks," shareholders and corporate CEOs who own and employ genetically designed "half-men" that are ferocious and powerful beasts, programmed for loyalty and obedience, i.e., slaves, custom-made to serve and die for their masters, since "only the richest corporations, the Chinese peacekeepers, and the armies in the North could afford to grow and use [them]."[8] Yet, life for the rich "swanks," as Nita Patel, heiress of Global Patel, explains, is equally violent and fraught with danger: "I am a chess piece ... a pawn ... [that] can be sacrificed, but cannot be captured ... [for then] they will have my father, and ... make him do terrible things."[9] Therefore, the earlier choice of either yielding to the powerbrokers' will or facing death is now replaced by the torture of others that makes yielding to the will of the aggressor even more compelling, for torturing others, one would like to believe, is a worse option than choosing one's own death.

The labor/energy the "swanks" appropriate from the half-men may be viewed as the energy equivalent of fossil fuels that are used and abused by predominantly by wealthy nations and has produced global warming and the current climate crisis. The half-men provide moral insulation to the "swanks" by doing their unethical and dirty work for them, chief amongst which is to terrorize, kill, and bend the poor people's will in service of the rich who view their duty to be the acquisition of profit and wealth. The corporate conglomerates, headed by "swanks," profit by fleecing people in all economic classes, including the lowest class of people, called the "licebiters," i.e., poor, disenfranchised children and adults, who like rag pickers, scavenge for reusable material, deploying their labor and diminutive size to break down complex, behemoth of ships, part by part, and element by element, for future capital gains. These societies that function under corporate management highlight capitalism at its worst, an economic system that compels people to compete against each other for basic necessities needed for survival. In *The Drowned Cities,* described as "a coastline swamped by rising sea levels and political hatreds, a place of shattered rubble and eternal gunfire," the fight for basic survival is played out in bloody colors.[10] This savage fight performs cultural misappropriation of human misconception of the law of the jungle that has constructed human identity as 'not animal' and positioned anthropogenic 'culture' as superior

---

[7] Ibid, 194.
[8] Bacigalupi, *Drowned Cities,* 71-72.
[9] Bacigalupi, *Ship Breaker,* 195-6
[10] Bacigalupi, *Tool of War,* 1.

to 'nature' in the nature/culture divide. Capitalism's championing of savage competition, along with its ideology of 'winner takes all,' glorifies brute force and strategy, rejecting rule of law and ethics that have been salient cornerstones of human culture. Furthermore, a capitalist economy makes consumption a goal by de facto, making the self-proclaimed exceptional species, Homo sapiens, nothing more than the sum total of its appetites, albeit the appetites of the "swanks." As Jason W. Moore argues in *Anthropocene or Capitalocene?* (2016), "Capitalism has thrived by mobilizing the work of nature as a whole; and to mobilize human work in configurations of 'paid' and 'unpaid' work by capturing the work/energies of the biosphere."[11] Moore explains:

> The rise of capitalism launched a new way of organizing nature. For the first time, a civilization mobilized a metric of wealth premised on labor rather than land productivity. This was the originary moment of today's fast fading Cheap Nature. The transition from land to labor productivity during the early modern era explains much of the revolutionary pace of early modern landscape transformation. The soil and the forests ... were appropriated (and exhausted) in the long seventeenth century. Human nature too was freely appropriated (and exhausted) as New World sugar frontiers and African slaving frontiers moved in tandem... these frontier-led appropriations were amplified by the fossil fuel boom. Fossil fuels were a new frontier.[12]

Bruno Latour's insight and emphasis on the complex and heterogeneous relationships between human and non-human agents and his likening of the production of scientific knowledge to the tracing of networks of relations among actors and materials that has come to be known as Actor-Network-Theory (ANT),[13] is useful in addressing the phenomena of anthropogenic climate change. Using James Lovelock's (1987) evocation of Earth as Gaia – an interconnected and interdependent network of living systems - and Bruno Latour's Actor Network Theory (ANT), it is possible to see how the promise of reason and science engendered during the Enlightenment was not quite the lightening rod it was chalked up to be. Although progress in the areas of science and technology has been significant, the integration of knowledge, necessary in mapping out the networks of relationships and

---

[11] Jason W. Moore, "The Rise of Cheap Nature" *Anthropocene or Capitalocene? Nature, History, and the Crisis of Capitalism,* ed. Jason W. Moore. (Oakland, CA: Kairos PM Press, 2016), 111.
[12] Moore, *Anthropocene or Capitalocene?* 110.
[13] Bruno Latour, *Reassembling the Social: An Introduction to Actor-Network-Theory* (Oxford, New York: Oxford University Press, 2005), 260.

agents amongst all fields of knowledge has not been pursued with due diligence. Consequently, as knowledge and understanding in the different disciplines grow in complexity, it becomes increasingly more challenging to visualize the ever-increasing webs of connectivity and material actants on Earth. However, no matter how challenging comprehending the complex network of actants is, it is not something that can be abandoned, for Earth's health and life itself depend on the understanding of such complex networks.

Donna Haraway's concept of the cyborg and understanding of Earth's environment as interplay of agencies in complex and complicated networks provides a potential paradigm to grapple with the notion of the Anthropocene and its effect on life.[14] By attending to actors, including non-human, material actors that New Materialism has introduced, a more inclusive and comprehensive conceptualization of Earth is made possible. The living conditions of the disenfranchised poor that multinational power brokers of the world choose to ignore, along with the injustices heaped on them, may unfold in ways that cannot be anticipated or managed. Moreover, the disenfranchised poor could be the very actants whose interactions with the environment could become the site for upending the dominant discourse of capitalism and expose its direct feedback loop to anthropogenic climate change. Because power brokers of the world have chosen not to attend to the disenfranchised poor and dumped their voluminous waste on land that the poor inhabit, they have not calculated the global impact of this population's interaction with its local, toxic environment.

In Bacigalupi's trilogy the space occupied by the disenfranchised poor is, also, the space occupied by damaged cyborgs called "half-men."[15] Because of the proximity of the disenfranchised poor and the rejected, or truant cyborgs, the two groups end up attending to each other, and develop an ecosystem of dependence that helps both groups to survive. For such groups, 'dependence' is not a value-ridden concept with pejorative connotations, but one that is necessary for existence and survival. Moore and other writers are mapping out underworlds, third, fourth, and fifth underworlds, occupied by 'actants' or agents whose sheer existence alters the materiality of Earth. For example, in The *Overstory* (2019), Richard Powers uncovers the marvel of Earth-wide network of tree roots and plant life that, although necessary for human existence and sustenance, often does not enter human consciousness. Because the underworld of human and non-human materiality has been ignored and the toxicity of the volatile subaltern

---

[14] Donna J. Haraway, *Simians, Cyborgs, and Women: The Reinvention of Nature*. London: Free Association Books, 1991.
[15] Bacigalupi, *Ship Breaker*, 45.

worlds unacknowledged, they are ripe for violent eruption that could throw light on Capitalism's appropriation of Earth materials and energies in its profit-making schemes and usher in a different kind of economic system that could place a premium on sustainable living and Earth restoration and preservation.

The cyborgs in Bacigalupi's trilogy are powerful, part human and part beast creatures, genetically programmed and trained to kill, terrorize, and obey. Yet, one half-man, Karta-Kul, takes on the moniker, "Tool," undoes his genetic programming through sheer will and effort, denies he is owned by anyone, and disobeys his masters, for which he is relentlessly persecuted. Nonetheless, against all odds, he survives and escapes into the abandoned Drowned Cities. On his disappearance, he is presumed dead, until news of a "half-man" leading an army of "licebiters," reaches the corporate world. But Tool, the tool, has better insight into the power wielders' operating system, than his designers their product; consequently, Tool knows exactly what his creators are thinking and, thus, is able to outsmart them. As Tool observes: "My creators do not fear my individual rebellion. They fear the uprising that I will eventually lead."[16] As feared, Tool ignites an uprising, defeats and, then, methodically unites warring warlords from the Drowned Cities, abandoned by the "swanks" that neither the peace-keeping forces from "civilized" China, nor the dreaded security companies could manage or subdue and, hence, abandoned. China, as Mahlia's father used to say, is civilized because "Chinese people didn't war among themselves. They planned and they built. When the sea rose, they built huge dikes to protect the coastlines and floated their great cities on the waves, like they did with Island Shanghai."[17]

Moreover, only after Tool escapes their imprisonment does the corporate conglomerate realize what a grave error it has made by manufacturing "half-men," as all members nod in unison at the emergency security meeting where R & D declares: "This technology must be terminated. It was a dangerous precedent and a foolish risk to try to create it. Caroa was a madman."[18] The power brokers' response to Tool's defiance is the elimination of defiance, and when he disappears, they make General Caroa the scapegoat for all that has gone wrong, when without the collusion of all the corporate players, especially the financiers, the production of half-men would not have been possible. They declare Caroa insane, forget about the security breach, and go on with their delusion of business as usual. The power brokers think that just because they, with their imagined power,

---

[16] Bacigalupi, *Tool*, 279.
[17] Bacigalupi, *Drowned Cities*, 61.
[18] Bacigalupi, *Tool*, 268.

decide to end manufacturing the product, all is well with the world and the problem of genetically altered half-men solved. Moreover, those who think that creating "half-men" was a bad idea do so, not out of a sense of wonder or humility at the unfathomable complexity of life at the most basic building block level, but because ownership of their tool/product has slipped their grasp and threatens their carefully constructed, but illusory power. Ownership of the production of the newest tool/product in the marketplace has enabled corporate brokers to exercise their power, but when the tool is lost, or beyond their control, their power is diminished; yet, they refuse to believe that it is their ownership of the tool that gives them their power, not their skill in tool making, or tool usage, as they delude themselves into believing. Humankind enamored by their tool-making ability forget their dependence on the tool for their perceived power and believe themselves to be the powerful ones, forgetting that they have used Earth's materials to create the tool, materials whose potential power/energy they have appropriated in manufacturing the tool. Nonetheless, so deluded are the corporate hustlers with their temporary power that they think they can maintain their power, forever, by maintaining the illusion of ownership of the tool that they have manufactured. By stopping production of the tool/product, they think they can put the genie back in the bottle without incurring any collateral damage; by acting in a decisive, powerful manner, they delude themselves into thinking that they are in charge. They declare both Tool and his maker, Caroa, dead, when in reality Tool has killed Caroa and disappeared into the vast debris of their material waste, in the uninhabitable seascape of Drowned Cities.

Jayant Patel, owner of Global Patel, has the clearest understanding of humans' inability to handle power when he declares to his daughter, Nita Patel: "Power poisons us, and it poisons them. It poisons so very much that I sometimes wish I had never made this company what it is."[19] The solution to corporate power's ability to poison and corrupt is either to dissipate power so that it does not accumulate in one location as it does in a capitalist society, where backstabbing, backdoor deal making, corporate mergers and hostile takeovers are the accepted and approved means of operating and profit making; or, for all Homo sapiens to realize in a deep, humble, and rational way that the power they appropriate from Earth's materials does not belong to them and hence can never be owned by them; they can only have temporary custody of the power of the tool in their hands. Tool, the half-man, whose nemesis is Homo sapiens, articulates the toxic relationship

---

[19] Ibid, 249.

between humans and power in another way. He declares: "Humans dislike weapons that think for themselves. It unnerves them."[20]

From Tool's characterization it becomes apparent that humanity's scientific productions, the tools, or cyborgs in this case, can and do transgress the boundaries of their programming and defy imposed limitations placed on them. Tool has much more discipline, self-control, and will power than all humans in the text. Although he has been designed by human agents, he escapes both his design and their manipulation, through self-understanding and clear observations, by means of which he evolves. He declares to Mahlia, "I am self-taught."[21] He is cognizant of his own strength and skills and, thus, is able to deploy them purposefully. He prefers to walk away from battles he knows he cannot win, unlike the armless human, Mahlia, who risks everything to save both Tool and her friend, Mouse, when neither of them are her "kin," or belong to her "pack."[22] Mahlia, who herself has gone through excruciating suffering, recognizes pain when she sees it in the powerful yet "patchwork Frankenstein,"[23] Tool, and gets him the care he needs to heal. For Tool, who has been used and abused as an agency-less tool by humans, being taken care of by a human is unfathomable. Consequently, he learns from this experience and others like it to care for other humans, like Mahlia, who are not his "kin," or "pack," demonstrating Haraway's notion of entangling multi-species oddkin.

When Tool, despite his DNA encoding, becomes a free, autonomous agent, he refuses to do humankind's reprehensible work, even at the cost of his own life. Tool feels indebted to Mahlia for her generosity, but even after he has repaid what he perceives to be his debt to her by returning the favor of saving her life, he cannot abandon her, even as she risks both their lives by attempting to locate and rescue her friend, Mouse. Tool traces a relationship with Mahlia that is not based on trade, exchange, repayment, or kinship. He admires the handless "war maggot's" generosity and tenacity which in turn softens the "slaughter bringer"[24] to, not only, spare the life of his worst enemy, Ariel Jones, but also, to rescue her. Furthermore, just when he thinks he is dead, he sees the "licebiter," Nailer, and the "swank," Nita, speeding to rescue him; he is overwhelmed by "humans working to save him," when he has always been used to take humans' deathblows.[25] He understands that not all humans are like General Caroa who trained him, or

---

[20] Ibid, 98.
[21] Ibid, 258.
[22] Ibid, 368.
[23] Ibid, 361
[24] Bacigalupi, *Drowned Cities*, 56.
[25] Bacigalupi, *Tool*, 368.

most other humans enculturated to both fear and despise him, and that, although few, there are other humans like Mahlia, Mouse, Nailer, and even "swank" Nita, who are not thoughtless, users and abusers of tools, and recognize that tools are made from Earth materials and hence have energy and agency of their own, derived partially or wholly from Earth materials. When Mahlia and Ocho find Tool severely injured and writhing in pain, Ocho who has seen too much pain and suffering thinks that ending Tool's pain and misery will be an act of kindness; however, Tool responds with utter affirmation of his agency: "I. Am. Not. *Meat*,"[26] i.e., he is not just a lump of flesh experiencing pain like humans do, and, thus, like humans, obsessed with ending suffering and resuming active consumption in the predator-prey chain of survival of the cruel dominator, but something different, and unlike humans in its needs, wants, and desires, differences humans cannot recognize or comprehend. So long have humans been told to recognize the humanity of their fellow humans to avoid discriminating and abusing them, that they cannot think in any other way, other than through the lens of human exceptionalism, obliterating the moment a new way of thinking could be ushered in, a way of thinking that could be inclusive of all life and all materials not like them, but different from them.

Tool also realizes that people like Mahlia and Nailer who act with compassion, and recognize the agency of what appears to be inanimate Earth materials, choose to do so, for compassion and recognition of material agency is not something that is hardwired into their DNA. Because Tool has been told that he has been constructed and programmed by humans to serve them, he believes he has no choice but to do so, especially kill and slaughter against his own will. Since Caroa and Mercier have created him, he believes they are his gods, but when he realizes he has bloodied his hands by doing their killings for them, he ends the two cowards' lives; perplexed he thinks, "*I slew my gods ... I am a God Slayer.*"[27] He kills Caroa and Mercier for the wrongs they have done him and feels no guilt for all the past slayings he has performed under orders and threats of annihilation. Caroa and Mercier deserve death for bloodying his hands to keep theirs' clean and for forcing him to believe in the constructed ideology of the savagery of existence, with its constructed reign of the will and ideology of the powerful. The power of the individual human is a constructed power that allows him to exercise his will and thus can be deconstructed too. When Tool, the augment, understands the construction of his oppression and dependency, he recognizes and understands the construct of his own agency, constructs it, and uses it.

---

[26] Ibid, 65.
[27] Bacigalupi, *Tool,* 340.

Furthermore, the power brokers would like all people to believe that the ideology of survival of the strongest and the most powerful, is the only true condition of life and those who refuse to believe in its truth and do not engage in its savagery are weak and foolish and hence will die prematurely. Although he is not a "swank," Richard Lopez, lives by the ideology of the savagery of survival that the corporate power brokers have adopted as a necessity or sacred mantra, and the manufactured half-men have been genetically wired to believe; yet, while Tool a half-man wishes he could go against the grain of his DNA hardwiring and training for violence and slaughter, Lopez freely chooses to embrace this ideology of predatory violence. Because Tool does not believe in the ideology of violent savagery as the condition of life, he wills himself into overriding his genetic hardwiring and war conditioning to embrace the ideology of generosity and compassion as he vehemently declares, "Blood is not destiny no matter what others may believe,"[28] meaning DNA, the building block of life, is not the only and final determinant for behavior and action, but that there is a whole, wide plethora of permutations and combinations he can will into existence or construct.

Early in the first book of Bacigalupi's trilogy, Nailer, the self-sufficient picaroon, ponders on the first and most basic construction of human culture: the institution of family. The only family Nailer has is his drug-addled father, Richard Lopez, who is described as "an apparition of evil,"[29] because all his life he has "been crazy and destructive ... and downright evil."[30] Lopez is Nailer's main torturer and keeps him in a state of perpetual fear. Nailer's musing on family leads him to the conclusion that, "most of those words and ideas [of family] just seemed like good excuses for people to behave badly and think they could get away with it."[31] At the end of the third book of the trilogy, there is a similar musing on family; this time it is the "half-man," Tool, who contemplates on the notion of family. Baffled by Nailer and Nita's attempt to rescue him, i.e., of a predator rescuing a prey, one species watching out for another, he settles on accepting kind and enlightened humans like Nita, Nailer, and Mahlia as "kin, if not in blood, then in kind. Pack."[32] For Tool, family, kin, and pack include all creatures that will disappear when they die, and hence present when alive, and thus deserving of all opportunities to live.

---

[28] Bacigalupi, *Ship Breaker,* 248.
[29] Ibid, 219.
[30] Ibid, 317.
[31] Ibid, 274.
[32] Ibid 368.

Although the half-men are called "devil's spawn"[33] by some, the term "evil"[34] is never used to describe them as it is used to describe Richard Lopez. In fact, despite their tremendous power and the fact that they are "faster, stronger, and ... smarter than ... [their] patrons,"[35] "People," says Tool, "always believe I am only their dog."[36] Compassionate, intelligent people intuitively perceive the presence of choice in all circumstances, and when people act with compassion they choose to do so, and when people do evil acts, they choose to do so too. Tool is genetically wired to cause pain and suffering, but Lopez chooses to addle his brain with drugs and alcohol and do cruel deeds. Tool has been programmed to spread maximum harm; yet, despite his genetic programming, he acknowledges good deeds and acts of kindness that he witnesses, learns from them, and enacts similar acts of kindness. Tool has never quite believed what he has been told about half-men. He tells Nita that he is "smarter than they prefer ... [and is] smart enough to know that I can *choose* who I serve and who I betray, which is more than can be said about the rest of my people."[37] Tool's articulation of the importance of choice and his stand on choice, separate him from other half-men. Later, when Tool confronts Ariel Jones, she, in an effort to save herself, says that she was just doing her duty and carrying out orders. Enraged by her excuses, Tool says: "There is *always* a choice ... You say you had no choice ... but that is not true ... My kind has no choice. We are *designed* to have no choice ... And. Yet. I *choose*."[38] Tool's insistence on choice is a sign of his evolution which begins with the observation of a compassionate act, followed by the emulation of the act, and finally to the deprogramming of his DNA for violence and harm.

If the developed world's current, worst nightmare is not perceived to be climate change but terrorist attacks that harm "innocent" people, i.e., people who have caused the terrorist no direct, personal harm, the misuse and abuse of knowledge that the production of knowledge makes possible has to be addressed. The productions of science and technology empower the rational and thoughtful as well as the irrational and thoughtless, and, hence, the producers of such knowledge bear some responsibility in the production of knowledge that the irrational use and abuse. Humans cannot produce weapons of mass destruction and be naïve or dumb enough to believe that it will not fall into the wrong hands, or will not be used for

---

[33] Bacigalupi, *Drowned Cities*, 5.
[34] Bacigalupi, *Ship Breaker*, 248.
[35] Ibid, 212.
[36] Ibid, 248.
[37] Ibid, 211.
[38] Bacigalupi, *Tool*, 362-3.

wrong ends. The US dropping of the atomic bomb on Hiroshima and Nagasaki may have ended World War II and brought an end to unspeakable suffering and destruction, but the knowledge and technology produced in the Manhattan Project is for sale for the right price; more importantly, the production of such knowledge also creates a desire for such knowledge in the market place, so much so that even the poorest countries in the world choose to devote more of their resources to the acquisition of such knowledge and technology than on basic necessities for their citizens. The capitalistic creation of desire for commodities along with the production of the commodities becomes a substantial challenge for the puny, individual, human will to combat. The flooding of cheap goods, including cheap food items in the marketplace for profit and gain is making large populations of people who consume these goods sick; the environmental pollution produced to manufacture these less than desirable items is eventually going to make the elite capitalist manufactures ill too, for neither pollution, climate change, nor pandemics respect human concocted, artificially imposed boundaries that separate and divide and keep the privileged and the disenfranchised in isolated enclaves.

While earlier, fictional monsters like Mary Shelley's Frankenstein monster craved human company, current monsters like Bacigalupi's Tool, find human company abhorrent and provide the most accurate yet devastating indictment of humankind: their willingness to do unto others what they themselves cannot bear. As Tool hears the boy, Mouse, beg for mercy in *The Drowned Cities,* he observes that the boy is "just another human who would grow up to be a monster [and] that humans ... [are] always ... so willing to do their worst to others, and always begging for mercy in the end."[39] When humans manipulate materials in their environment with partial or incomplete knowledge, without considering the agency of the materials and their connections to networks of actants, they are acting foolishly and playing with unknown fires that could potentially destroy everything, including themselves.

Although Tool is not a terrorist, he may be viewed as the ultimate terrorist - a perfected, killing machine, designed to produce maximum harm. The network of actants that constitute the perfected killing machine are the materials used, the tool maker, the tool, and the tool user all fused into one singular entity whose component parts cannot be separated from each other, or from the sphere of their operation. Tool's masters want him to remain a tool and thus separate him and keep him away from their own selves so that their imbrications in all parts of his existence is rendered invisible.

---

[39] Bacigalupi, *Drowned Cities,* 139.

However, such separation even if it is imposed is not possible, because the agency of the tool and other networks of connections have only been veiled for human manipulation without accountability. Tool is a tool, but how this tool is affected and used is not just determined by the tool maker or tool user, but, also, by the tool and the network of actants in its web of connections. Bacigalupi, by constructing Tool as half man and half beast, humanizes a tool of war and brings to life the network of the tool in its environmental agency, which hitherto has been dismissed as inanimate material with no power or agency.

Furthermore, terrible as the half beasts are made out to be, they are not as bad as humans, such as Nita's uncle, Pyce, or Nailer's father, Richard Lopez, in the way they treat their kin. When Lopez is burning "with amphetamines and liquor," he thinks of the "maps the violence"[40] he will etch out on materials and people within his sphere of operation. That Lopez through his insatiable consumption of alcohol and drugs is more violent to his son than a half-man programmed for maximum harm is indicative of the violence built into the capitalistic system with its insatiable appetite for commodities and tools. In developing Lopez's character, the narrator says, "Drugs whittled him [Lopez] down to a burning core of violence and hunger."[41] Although he is poor, Lopez's unchecked consumption is the cause of his violence, suggesting, perhaps, that unchecked and unregulated consumption, ultimately leads to violence, and that consumption, capitalism, and violence all feed into each other to emerge into existence. Commodities' potential for violence is further elaborated in the trapped condition Nita is found in, when her swanky ship crashes into Bright Sands Beach. Nailer and Pima discover Nita "pinned under the pile of her bed and the weight of all her *stuff* that had crushed her"[42] (emphasis added). When the two realize she is alive and try to rescue her, it takes them "almost an hour to pull all the stuff free,"[43] so that they can unpin her from under the weight of all her belongings that weigh down on her. All of Nita's belongings jeopardize her safety and delay her rescue, again suggesting the destructive potential of stuff consumer capitalism compels people to own, even when they are unnecessary and destructive to their health and life.

---

[40] Bacigalupi, *Ship Breaker,* 55.
[41] Ibid, 58.
[42] Ibid, 90.
[43] Bacigalupi, *Ship Breaker*, 101.

Conclusion

The current coronavirus pandemic is, perhaps, an expression of the unattended non-human material life. A virus, whose diminutive size makes it invisible to the naked eye, by expressing itself on a host, has brought the Homo sapiens species to its knees. If we wish to survive, live, and express ourselves, collectively as a species and individually as our own selves, we must attend to all human and non-human material around us - those with energy and those without energy that can ride on the life and energy of others and cause havoc, the likes of which we have not seen as yet.

Our ancestors and indigenous cultures that still live in close proximity to the natural world, without creating an excess of artificial boundaries around them, are more aware of the material world and its power. When the Europeans colonized the world, they decimated these ancient cultures and what they didn't decimate they made illegitimate and undesirable. A major tool of colonial power deployed in India by imperial Britain was the Christian missionary whose goal was to bring natives into the Christian fold. To achieve their mission, these missionaries delegitimized indigent religions and practices, denigrating what they termed 'idol worship' and India's 'many gods,' advocating for the worship of one god, the Christian God. Little did the British missionaries realize that Hinduism was never deemed as a religion by its practitioners but as a way of life where active and non-active material life's power and presence were symbolically acknowledged. In India, even today, stones, trees, animals, plague carrying mice, and venomous snakes are all revered and ritually worshipped, which may appear as foolish superstition but viewed through ANT is an acknowledgement of all materials, species, and networks no matter how vile and pernicious they may appear to humans. Indigenous cultures living in close proximity to material Earth, revere mountains, rivers, landscapes, etc., and perceive them as being imbued with spiritual life. When viewed through ANT theory these beliefs and practices may be interpreted as an acknowledgement through ritual worship the materiality and network of actants of these Earth structures. Because the human mind cannot hold the presence of multifarious materials and non-human life all at the same time, it is incumbent on humans to develop a sense of humility to acknowledge our own miniscule significance in the vastness of all there is to know and discover about Earth and the universe. How we do so is through sheer attention and observation of all that is culturally being made irrelevant through globalization, where making quick and fast profit is the be all and end all of existence.

Thus, the more humans try to manipulate their material environment, the more agency the material environment with its network of actants reveals. Humans can try various methods of manipulating the nonhuman material world, or appropriate its material energy, but they can do so only temporarily, for in the long run, the non-human material agents regain their actant equilibrium and strike back at human culture and society with vengeance. When viewed through ANT theory, this response of material Earth may be viewed as the human actors' recognition of networks, or material actants' tracings, hitherto unnoticed, ignored, or dismissed by humans. The power brokers who manufacture "half-men" and General Coroa who trains them are convinced of their total power and control over Tool; however, the sheer arrogance of such thinking and the unintelligent choices of these megalomaniacs are revealed, not only by the disastrous failures of their Earth altering endeavors, but also by the long term harm and havoc they unleash. The western world's knowledge of science and technology has made life comfortable and far better for a vast number of people; however, the questions that are not being asked but must be asked during such enhancements are at what cost and for how long these enhancements will last.

References

Bacigalupi, Paolo. *Ship Breaker*. New York; Boston: Little Brown, 2010.
—. *The Drowned Cities*. New York; Boston: Little Brown, 2012.
—. *Tool of War*. New York; Boston: Little Brown, 2017.
Coleridge, Samuel Taylor 1834. "The Rime of the Ancient Mariner." *Poetry Foundation*. Poetry Foundation. Org.
Haraway, Donna. *Simians, Cyborgs and Women: The Reinvention of Nature*. London: Free Association Books, 1991.
Jahren, Hope. *The Story of More: How We Got to Climate Change and Where to Go From Here*. New York: Vintage, Random House, 2020.
Jennings, Hope. 2019. *Contemporary Women's Writing*. 13.1: 16-33.
Kolbert, Elizabeth. *The Sixth Extinction: An Unnatural History*. New York: Picador. Henry Holt Company, 2014.
Latour, Bruno. *Reassembling the Social: An Introduction to Actor-Network-Theory*. Oxford, New York: Oxford University Press, 2005.
Martel, Yann. *Life of Pi*. Orlando; Austin; New York: Harvest, Hardcourt, 2001.
Moore, Jason W. "The Rise of Cheap Nature" Anthropocene *or Capitalocence: Nature, History, and the Crisis of Capitalocence*. Ed. by. Jason W. Moore. Oakland, CA: Kairos PM Press, 2016.

Nixon, Rob. *Slow Violence and the Environmentalism of the Poor.* Cambridge, Massachusetts; London, England: Harvard UP, 2011.
Powers, Richard. *The Overstory.* New York; London: W.W. Norton, 2019.
Oreskes, Naomi and Conway, Erik M. *The Collapse of Western Civilization: A View from the Future.* New York: Columbia UP, 2014.
Shakespeare, William. *King Lear. William Shakespeare*: *The Complete Works.* 5th Hardcover. Edited by David Bevington, 2003.

CHAPTER TWELVE

REVELATIONS OF THE MILLENNIUM: POSTHUMAN BEINGS AND/ IN POSTAPOCALYPTIC WORLDS

ANDREW ERICKSON

Introduction

Millennialist projections of world destruction and its transformations of peoples and worlds continue to persist in the US-American imagination of apocalypse. The earliest European settlers took their reading of the biblical Revelation seriously and declared themselves the people revealed to have been chosen to enlighten the "new world" and inherit its millennial kingdom. Since then, this idea of literally embodying the metaphorical eschaton took another radical turn in the twentieth century, as new constellations of millennialists formed to project all manner of world-endings and renewals in coalitions across lines of class and race. This diversification arose, in part, as the result of a revelation that human technologies had made possible the actual end of the world (as witnessed in the German genocide of Jewish people and the U.S. bombing of Hiroshima and Nagasaki), and also from a growing cultural awareness among white Americans that, for some within the United States, this was also already true, as Black and Indigenous peoples and People of Color lived on through the apocalypses of colonialism and enslavement. My contribution shows how millennialist thinking about the apocalypse in this vein structures attitudes toward environmental "world" destruction that consider the literal end of the world as an historic event that has already happened. As the growing climate emergency takes center stage in the twenty-first century, we are perhaps now hyper-aware of this, yet misunderstand what it means for those "on the ground." Jesmyn Ward in *Salvage the Bones* (2011) reorients millennialist thinking toward how catastrophe continues to be characterized as an apocalyptic event and is survived. She does so by considering what it means to live reconfigured, postapocalyptic lives in the millennial aftermath of "the end of the world." In her novel this takes the form of a fictionalized Hurricane Katrina that literally destroys the world of its characters, echoing the real-life destruction that Ward herself experienced in her community. My contribution thinks together millennialist and postapocalyptic worldviews to consider how Ward's novel reconfigures millennialist apocalypse in terms of peoples who really live through catastrophe, and also the worlds they inherit.

Ward engages a millennialist apocalypse that reconfigures subjectivities of an impoverished Black family (including animal companions) at the center of her novel. In this she problematizes often-repeated millennialist thinking, which follows the biblical Revelation to prophesy an apocalypse that ends one world as it begins another, in which lives are radically transformed. Often such prophesies have placed undue emphasis on "chosen" peoples; Ward shifts the scope of inclusion. Those dehumanized in an hegemonic American history and in traditional millennialism take agency in the novel. *Salvage the Bones* rethinks the postapocalyptic subject as it locates humanity within the natural world in what I read as a posthumanist turn. Ward situates this turn within contemporary thinking about apocalypse and millennium, as she reorients the (post)apocalypse to indicate how those most affected by literal world destruction are already living on—literally and figuratively—through the recurring storm.

What the Apocalypse Reveals

Apocalypse in the biblical Revelation tells of the end of the world and an aftermath in which a formerly oppressed people are chosen to inherit a millennial kingdom. This forms the core of popular millennialist belief. In its guarantee of future for a chosen people, it remains a powerful narrative within U.S. culture. Fukuyama famously engages apocalyptic language to evoke a teleological "end of history" as humanity is projected to reach its predetermined, final state of secular enlightenment.[1] Religious fundamentalists read apocalypse in the traumatic events of American history. Early Puritan settlers, like Columbus before them, engaged in a (re)telling of their removal to America as an apocalyptic event for which they had been

---

[1] Francis Fukuyama, *The End of History and the Last Man* (New York: The Free Press, 1992), 56.

divinely chosen, leaving them to inherit its "millennial kingdom."[2] Charles B. Strozier indicates the prevalence of apocalyptic language used to describe enslavement and the U.S. Civil War that fought to end it, noting that mid- to late-nineteenth century "gave reality to the images of apocalyptic horror."[3] This split in apocalyptic thinking engenders the end as metaphoric redemption and as a metonymic condition of life, and is accommodated in a millennialist understanding of world destruction. Strozier notes that millennialist fundamentalism follows from a literal reading of the biblical Revelation to John, which includes

> a violent and destructive end of human history, followed by some kind of parenthesis that fundamentalists read as the millennium, and the great climax of the final judgment which sorts out those who go to heaven while nonbelievers are thrown into the lake of fire. The apocalyptic, in other words, relates the specific forms of our forthcoming destruction.[4]

He goes on to argue that reading in both modes enables a plurality within apocalyptic thinking, which explains, in part, its popularity among such a wide and deeply differentiated segmentation of the U.S. population.[5] Daniel Wojcik also points out that millennialist belief enables thinking apocalypse in both a literal and a figurative mode: "true believers" imagine a divine revelation in the events of their time, including cataclysmic transformation that would "eliminate evil and establish a terrestrial paradise" and "usher in the millennium."[6] Wojcik and Strozier confirm the broad appeal and application of this central idea of a divine plan for an end to history that is revealed to a chosen people who will then inherit the millennial kingdom in the aftermath of apocalypse. Such reification of the apocalypse as a real historic event increases significantly with the world wars of the twentieth century. In particular, telling of the Second World War as apocalypse creates an aftermath in which projections of the hopelessness and despair of the present age and the urge to usher in a new one reached a

---

[2] Daniel Wojcik, *The End of the World as We Know It: Faith, Fatalism, and Apocalypse in America* (New York: New York University Press, 1997), 21-23.; cf. Sacvan Bercovitch, *The American Jeremiad* (Madison: University of Wisconsin Press, 2011 [1978]) and Sacvan Bercovitch, *Puritan Origins of the American Self* (New Haven: Yale University Press, 2011 [1975]).
[3] Charles B. Strozier, *Apocalypse: on the Psychology of Fundamentalism in America* (Boston, MA: Beacon Press, 1994),172.
[4] Ibid., 154.
[5] Ibid., 154.
[6] Wojcik, *The End of the World as We Know It*, 13.

fever pitch.[7] The extermination of millions of people alongside the development of enhanced necrotechnologies seemed to harbinger the end of the world; or of humankind. As globalization reorganized the movement of peoples and their modes of being in the world, the same suggested to some that an end to human development (of its world) had finally arrived.

Millennialist apocalypse maps the atrocities committed throughout the twentieth century. Surveys of fundamentalist eschatological thinking in this period, however, indicate that a fundamentalist reading of Revelation projects too much into a future that may never come to pass, forgetting in the process a past that has actually happened. A fundamentalist reading of Revelation relates "one or more revelatory visions about the future or the heavenly world or both" to "reflect a sharp distinction between the present evil age and the imminent future age of blessing."[8] In Revelation the narrator, John, establishes a frame in which God reveals to him a divine plan of coming wrath for an unrepentant world. This suggests *the* apocalypse (singular) as a temporal moment very much *to come* (a future singularity). Millennialists take *fait accompli* this rendering of apocalypse. As Strozier notes,

> fundamentalists' broken narratives profoundly distort time, a break that is rooted in experience. The past is separated off, to be remembered only as an abject lesson …. Things can be delayed but not solved. There is no redemption in the human purpose. Culture is rotten. The only hope lies in the mythical transformation of the future, in the remaking of the world during the millennium and the ultimate salvation in heaven after the final judgment.[9]

Thus in a literal reading the apocalypse looms just beyond the horizon. An isolated "mythical transformation" disrupts the past-present-future continuum in an event that functions something like a pressure relief valve: John specifies that a chosen few will be saved, raptured out of the despairing world, and brought into an utopian heavenly kingdom. A dramatic *deus ex machina* absolves a chosen few of their responsibility toward the human and nonhuman lives left behind by the apocalypse, and subsequently frees these "chosen people" from the consequences of the past they share with those who remain unchosen. Such a projection of world

---

[7] Ibid., 28.
[8] David Aune, "Revelation: Introduction." In *The Harper Collins Study Bible: New Revised Standard Version*, ed. Wayne A. Meeks (New York: Harper Collins, 1993), 2309.
[9] Strozier, *Apocalypse,* 45.

destruction forgets the atrocious historical moments of conception from which it derives. Literal insinuations of chaos and totalizing destruction in Revelation situate redemption in an unknowable future.

The question of who might be chosen and how further problematizes this discourse, as the apocalyptic narrative has been used to serve many different masters in the late twentieth and early twenty-first centuries. In the 1970s charismatic religious leader Jim Jones convinced his followers that the world was corrupt and demeaning to true believers of his teachings; only cataclysm (through proscribed mass suicide of his followers) would bring about a "socialist millennial kingdom" for them on Earth.[10] This applies a brutally literal logic of surviving the end, which unfortunately was not born out, and led to the unnecessary, yet seemingly voluntary, death of 913 people.[11] Engaging this rhetoric of chosen people to support an alternative agenda does not exist only as the prerogative of cult leaders, but also in mainstream U.S. political and religious rhetoric. Ronald Reagan similarly gestured toward a divine plan that included justification for nuclear war to bring about the end of the evils in the world in a 1964 speech.[12] As U.S. President he continued throughout the 1980s to assure Americans that should such nuclear apocalypse come to pass, it was sure to usher in a millennial age of preordained peace and justice and overcome the evils of Communism.[13] In both instances he implied that apocalyptic destruction would somehow make the world safe for Americans and American democratic ideals, which in itself was not a new concept at the time.[14] Populist evangelical leader (and bestselling writer) Hal Lindsey professed a conviction that the "symbolic end time rhetoric in the book of Revelation" could be linked with the "more immediate political world [of the 1970s], the cold war, and the threat of nuclear annihilation."[15] These samplings of apocalyptic rhetoric indicate the many uses of the chosen people trope and its place in millennialist thinking. From colonization onward, it remains a

---

[10] Wojcik, *The End of the World as We Know It*, 29.
[11] "Jim Jones," Encyclopædia Britannica, accessed March 23, 2021. https://www.britannica.com/biography/Jim-Jones.
[12] James Berger, *After the End: Representations of Post-apocalypse* (Minneapolis: University of Minnesota Press,1999), 136-37.
[13] Wojcik, *The End of the World as We Know It*, 30.
[14] Woodrow Wilson. "The Ideals of America." In*The American Intellectual Tradition*, vol. II, fourth ed., ed. David A. Hollinger and Charles Capper (Oxford: Oxford University Press, 2001 [1902]), 124-25. U.S. President Woodrow Wilson engaged the same rhetoric in his justification for U.S. imperialism by way of participating in the World Wars to ensure that the world was made suitable for democratic ideals and, by extension, for the continued existence of "Americans."
[15] Strozier, *Apocalypse*, 54.

persuasive dialectic for American imaginers of the coming end and their implied chosenness. It suggests that its elect will continue on, despite all odds (and evidence) to the contrary; in this vein, millennialism also indicates a limited futurity. I now turn to locating the subject of that futurity in terms of those who literally live through the apocalypse to witness its aftermath.

Millennialist apocalyptic thinking from Revelation onward projects a future dispensation of justice that forms a new world order. While Revelation identifies the target of oppression as a minority subculture within its historical context,[16] certain millennialists popularize the notion that the chosen people in Revelation actually link metaphorically the Christian minority persecuted in the early part of the current era with the cultural majority of evangelical Christians living in the United States in the 1950s onward. Strozier identifies the irony of such misreading: while fundamentalist millennialism in practice links together people from diverse backgrounds, the epistemological structures of reading Revelation with literal weight for the present moment espouses an inherent racism:

> The difference between self and others is sacralized, which provides a totalistic framework for dismissing all those who do not fit into the holy world in which the fundamentalist is blessed. The fundamentalist's chosenness defines the nonbeliever's abandonment, the one's salvation the other's punishment. Nonbelievers are rejected by God and thus in some inexplicable way are only tentatively human."[17]

The situation of chosen and not chosen people, true believers and false ones (or even nonbelievers), imagines a normalized center and its othered periphery. Ward problematizes the promise that those chosen will be saved from destruction and enter unharmed into the millennial kingdom. In *Salvage the Bones*, there is no acknowledged outside to the narrative space that is threatened by climate catastrophe, and all beings are subject to the storm's world-destruction. The absence of chosenness reveals a tentative humanity that is shared as those who survive must endure the reconfigured landscapes of their mode of being in the world. In the "we" of the novel's narrator there are only "smaller animals." In this way the novel directly inverts fundamentalist reading of biblical apocalypse: where fundamentalist reading places emphasis on chosenness and inheritance and erases its others,

---

[16] Aune, "Revelation: Introduction," 2308. Aune notes that *Revelation* describes the persecution of Christians as a cultural minority living under Roman imperial occupation in the early part of the first century of the current era.
[17] Strozier, *Apocalypse*, 90.

Ward acknowledges those marginalized, "tentatively human" beings as the center. It is left to the reader to imagine what relatively "bigger animals" might exist outside the novel's narrative space. Perhaps Ward intends that her reader should understand that, in the end, none actually exist. In the end, the novel seems to say, those who believed themselves to be chosen, who hoped for a different future, will nevertheless face apocalypse again. These are not the "bigger animals" that provide the absent referent in *Bones*. There are only postapocalyptic peoples to be transformed by the end of the world.

Those most oppressed in U.S. history are often excluded from dominant narratives of the elect of history. Paul Gilroy shows in *Against Race: Imaging Political Culture Beyond the Colour Line* (2000) that legacies resulting from colonization and extreme capitalism have hardened racialized projections of who is included in dominant definitions of the "human" subject.[18] For many who continued to be dehumanized and excluded from power, this revealed the familiar context of Revelation in their own lives as they struggled with hopelessness and despair and dreamed of a renewal of the world that would leave them better off. *Salvage the Bones* describes apocalypse in terms of this precarity of Black life in the American South. In doing so the novel clearly indicates a millennialism that denies a hopeful or nihilistic mode. For Ward's characters, the apocalyptic event and its promised aftermath is not something to be hoped for or avoided, but rather structures the very conditions of life; apocalypse just is. And it projects an aftermath.

Many millennialist imaginers of apocalypse are marked by ethnicity and class that have traditionally been the subject of exploitation. Strozier notes that, although mainstream millennialism considers white European Americans to be its chosen people, its narrative of imagined futures and people who survive catastrophe has been expropriated by diverse cultural groups, including those generational descendants of the enslaved and colonized.[19] This can be attributed to the fact that, for many who have been historically excluded from discourses of power, the apocalypse as world-ending has played a very real role. In his survey of the postapocalyptic James Berger emphasizes lived world-endings and people who survive them. He calls the racialization and enslavement of human beings "the apocalypse in history that we are living after and that symptomatically permeates [American] culture."[20] The historic event carries meaning for those who live through its moment of destruction, bringing into view a

---

[18] Paul Gilroy, *Against Race: Imaging Political Culture Beyond the Colour* (Cambridge: Harvard University Press).
[19] Strozier, *Apocalypse,* 154-55.
[20] Berger, *After the End*, xiv.

usable past that structures the present. Berger engages the term "post-apocalypse" to refer to this phenomenon, in which the apocalypse paradoxically means an end of the world, but somehow leaves behind survivors. The paradox reveals the problem of representing the end: how can destruction be both totalizing and survivable? When the apocalyptic event is realized in history, how does the notion of "chosenness" fall apart? Ward addresses these questions head-on as a fictionalized Hurricane Katrina provides the apocalypse that affects everyone in its world. The millennium exists in the aftermath of the storm, which Ward reduces to a single, final chapter in the novel—a coda of sorts. It also exists in the ways in which the characters live postapocalyptic lives, remembering the destruction of past storms and their loss.

Postapocalyptic Sensibility

The problem of representing apocalypse from the perspective of those who have lived through the end of their world centers in part on the authority of witness. *Salvage the Bones* follows an impoverished Black family in the rural Mississippi delta as they anticipate and live through the storm. Repeated references to the bare life of the characters evidence the precarity that Ward tells an interviewer she wanted to capture for her reader: "[T]his is the truth. You know, like this is the reality for so many people where I come from and it was the reality for me for a portion of my life. [...] Like, some of the aspects of this, of the poverty that these characters experience, like, that's real. And I can't deny telling that truth."[21] Ward identifies herself as the bearer of the truths she tells in her novel, lending the weight of verisimilitude to the fiction. She undoubtedly justifies her depiction of precarious life in part to convey the truth of her own witness statement: I have been there, and this really happens. This indicates the difficult position of the witness as survivor of world destruction, and of a breakdown in representation thereof.

Despite the attempted and executed destruction of worlds, Black Americans as postapocalyptic peoples persist. The world ends and yet it continues in a paradoxical turn, and it is in this turning that Berger establishes the post-apocalyptic:

> Events have consequences, there are remainders to every catastrophe, and "obliteration" is always a relative term, for a cultural memory has many storage areas and modes of expression.

---

[21] Jesmyn Ward, "In Salvage The Bones, Family's Story Of Survival," *Tell Me More: Books*, NPR, December 5, 2011.

> To see a world as post-apocalyptic is to recognize its formative catastrophes and their symptoms, and to identify the ideological sutures that hide the damages and repetitions. It is also, finally, to recognize and create narratives that work through these symptoms to return to the apocalyptic moments that traumatize and reveal. At that point, new—more healthy and more truthful—histories and futures may be possible.[22]

Whether or not such a Freudian "working through" is possible or even applicable to Black lives in the aftermath of a devastating hurricane in real life, I engage Berger's concept here because I think it speaks to the point that Ward makes in her novel. Her presentation of Black life is, among other things, a direct statement about the "ideological structures that hide the damages and repetitions" of socioeconomic treatment of Black people in the American South. It might well present her effort to work through the postapocalypse to produce histories and futures for her characters. This can be read as an attempt to speak the unspeakable, and to insist on futurity against extreme odds. Berger establishes the (post)apocalypse as an event that is also not an event in the sense that it consists of a moment of traumatic horror that can somehow not be spoken and so fails the present consciousness, yet insists on its own existence as something that can be witnessed: "In the late twentieth century, the unimaginable, the unspeakable, has already happened, and continues to happen. And, paradoxically, while unimaginable, it is at the same time quite visible."[23] Speaking truth of the impossible, yet already-having-occurred (thus possible), apocalyptic event remembers its cyclical trauma. This functions, in effect, as the "bubble collision" of two competing universes of truth.[24] *Salvage the Bones* speaks of the impossible made possible, of surviving against extreme odds to live out a future. To echo Michel Martin's comments during an interview with Ward, "it also tells the story of what they were enduring before the storm. Extreme poverty, sexual abuse, routine violence and survival."[25] The characters in *Bones*, impossibly, speak about their very real postapocalyptic lives with the weight of witness. But how can one speak about the end of the world when it ceases to remain an abstract prophesy and instead takes the literal coming storm that will surely obliterate everything in its wake, as its forebears have done? Representation proves impossible both because of the nature of the thing to be represented and because of the subject who

---

[22] Berger, *After the End*, 218-19.
[23] Ibid., 42.
[24] Cf. Timothy Morton, *Humankind: Solidarity with Nonhuman People* (London: Verso, 2017), 16-18.
[25] Ward, "In Salvage The Bones, Family's Story Of Survival."

attempts to speak of the unspeakable. On both accounts, this effort echoes the millennial revelation in the eponymous biblical book that the New Jerusalem to be established proves incommensurable for humankind—literally no earthly tool exists that can be used to measure it—and must be undertaken only by the angel with a divine rod of gold.[26] Measuring the incommensurate, in this regard, proves impossible without acknowledging the radically altered subjectivity of the subject; like the divine in Revelation, the subjectivity at the center of *Bones* is entirely other.

One way of thinking of the radical transformation of the apocalypse as historical event would be to consider its development into cultural memory. Berger conceptualizes apocalypse-as-historic event as a kind of trauma and a Freudian "working through" the repeated trauma as an engagement of the postapocalyptic subject with the moment of destruction.[27] This focuses on the ways in which individual experiences of apocalypse move into the communal domain. In one sense, the survivor of apocalypse *alone* bears witness to the moment of collapse, and the figure of the of witness calls to mind an image of a lonely individual, spotlighted in a dark room, attempting to speak of the unspeakable. In *Salvage the Bones* it most often falls to Esch's mother to provide witness of past destruction.[28] It is telling that the mother remains unnamed in the novel; as a dead witness who speaks of the unspeakable, she is rendered beyond reproach and her authority left intact, but like the angel in Revelation she exists as a superhuman entity who cannot be known. Her relative isolation is broken only when her family members re-member her to retrieve her witness (as wisdom) of past catastrophe. The dead mother comprises one such witness. "Daddy" Claude Batiste, the widower patriarch of the family, is another. He continually reminds his family that this part of the world knows plenty about what to expect about the end of the world, for they have already seen it in the double bind of being Black in the American South and in the return of tropical storms and hurricanes where they live.[29] For Ward the apocalypse takes the form of extreme weather that leaves the family patriarch continually looking for shelter from the storm in an effort to protect his family: "Daddy's crazy, I think, obsessed with hurricanes this summer. He was convinced last summer after one tornado touched down at a shopping complex in Germaine that the Gulf Coast would be a new tornado alley. He

---

[26] Rev. 21.15-17 (*The Harper Collins Study Bible: New Revised Standard Version*. New York: Harper Collins, 1993).
[27] Berger, *After the End*, 23-24.
[28] Jesmyn Ward, *Salvage the Bones* (New York and London: Bloomsbury, 2017), chap. 11, Apple Books.
[29] Ibid., chap. 10.

spent the entire summer pointing out the safest places in the house to crouch."[30]

In the way that he seems to be looking always for a way to survive another catastrophe, Daddy's behavior can be read as postapocalyptic, and also post-traumatic. Berger partially defines apocalypse as "historical trauma—a catastrophic and obliterating event that has occurred and that generates symptoms that define the world that follows."[31] The death of Esch's mother and past hurricanes loom as traumatic historical moments that anticipate the storm to come and reveal the anxiety of the characters in the wake of repeated and ongoing destruction. The past comes back again and again in a cyclical view of historical trauma. Ward acknowledges the cycle of repeated trauma in producing postapocalyptic peoples, and in doing so, she also indicates a sense of working or living through to a future. The singular event of the storm as apocalypse centers thinking about past cycles and their repetitions, and it forms localized knowledge that can be witnessed (sometimes in spectral and post-traumatic voices). In this way the moment of apocalypse becomes measurable for the radically transformed characters who live through it. The unchosen who endure the apocalypse live on to speak with the authority of witness as postapocalyptic peoples.

The scope of inclusion in the "chosen/unchosen" dichotomy provides the central conceit of much millennialist apocalyptic thinking. Strozier notes the positive assumption among millennial believers that they are the chosen ones, yet Ward reveals the unlikeliness of this being true. Esch recalls a conversation with her mother about survival in which they are revealed to be "other animals": "When Mama first explained to me what a hurricane was, I thought that all the animals ran away, that they fled the storms before they came. [...] And maybe the bigger animals do. But now I think that *other animals* ... don't do that at all. Maybe the small don't run. Maybe the small pause on their branches, the pine-lined earth, nose up, catch that coming storm air that would smell like salt to them, like salt and clean burning fire, and they prepare *like us*."[32] Esch describes the precarious position of the unchosen in terms of shared status as smaller animals, yet there exist no "bigger animals" in the novel. As the aftermath of the storm reveals, even the wealthiest face destruction: "the old white-columned homes that faced the beach, that made us feel small and dirty and poorer than ever when we came here [...] are gone. Not ravaged, not rubble, but completely gone."[33] The particular formulation of the novel's people as

---

[30] Ibid., chap. 3.
[31] Berger, *After the End*, 59.
[32] Ward, *Salvage the Bones,* chap. 11 (emphasis added).
[33] Ibid., chap. 12.

animals reveals a mode of being in the novel that departs substantially from earlier configurations of chosen people and their millennial kingdom. In place of this earlier iteration of the human stands a radically altered postapocalyptic subject.

The Posthuman as Radical Subject(ivity)

Another way of thinking through the radical transformation of the postapocalyptic subject would be to think of the humanity at the center of Ward in entirely other terms of its subjectivities. *Bones* uses language that clearly connects its characters in a shared ontological subject position to evacuate the human-animal binary and evidence an*other* mode of being entirely. This echoes the call of posthumanism to move beyond the humanism of Enlightenment thinkers toward a more inclusive human being. There are not chosen and unchosen peoples, but instead "other animals" who prepare "like us." Rosi Braidotti in *The Posthuman* (2013) argues for "critical posthumanism" to restore and center the "others" that European colonization dehumanized. Primarily this means combating "the reduction to sub-human status," where this reduction constitutes a "source of ignorance, falsity, and bad consciousness" on the part of the "dominant subject who is responsible for their epistemic as well as social dehumanization."[34] The inclusion of human and nonhuman actors in the formulation of "smaller animals"—as well as the novel's outright repudiation of "bigger animals"—seems to answer Braidotti's call to "devise new social, ethical, and discursive schemes of subject formation to match the profound transformations we are undergoing."[35] Where posthumanists like Braidotti call for this creature to come into being, however, Ward shows that she already exists. As a result, I wish to evaluate posthumanism as a politics of becoming that considers the emergence of new subjectivities in *Salvage the Bones*.

---

[34] Rosi Braidotti, *The Posthuman* (Cambridge: Polity, 2013), 28.
[35] Ibid., 12.

The core of posthumanist thinking comprises an evacuation of the human-animal binary opposition and a mutual recognition of agency,[36] [37] [38] which humanism has historically denied its others. Timothy Morton notes that speciesism and racism, in this regard, have both been used as tools of white elites to effectively diminish the human status of Black people, as well as animals. He situates this in the context of the caste system that "distinguishes between humans; then the distinction is mapped onto nonhumans. The tendency to see nonhumans as unthinking and even unfeeling machines is predicated on the objectification and dehumanization of other humans."[39] Zakiyyah Iman Jackson further notes "discourses on 'the animal' and 'the black' were conjoined and are now mutually reinforcing narratives in the traveling racializations of the globalizing West."[40] While *Bones* does not explicitly comment on the racialization of her characters, it nevertheless indicates how Black life imbricates the lives of nonhuman others. The novel acknowledges the limited agency of the family's pit bull, China. In one sense China is reduced to animal capital: Skeetah breeds her out to improve the family's poor living conditions and send his brother on to a better life,[41] and he also forces her to fight other

---

[36] Donna Haraway, *Manifestly Haraway* (Minneapolis: University of Minnesota Press, 2016). Haraway's figure of the cyborg (and perhaps also of Dolly the bioengineered Sheep) remains an early example of how thinking human ontology in terms other than Cartesian minds disconnected from bodies becomes possible, instead focusing on the embeddedness of humans in their environments; Cary Wolfe, *What is Posthumanism?* (Minneapolis: University of Minnesota Press, 2010).
[37] Donna Haraway, *Staying With the Trouble: Making Kin in the Chthulucene* (Durham: Duke University Press), 101-03. In her later work, Haraway addresses the problematic image of the cyborg as hybrid (neither wholly "human" or other, thus altogether alien), and compliments/supersedes it with the concept of "companion species" and "making kin," in which human (or "enlightenment Man") and other-than-human relations are consented to and continuously created in symbiotic networks of making and "becoming-with." It also bears keeping in mind that Haraway refuses the term "post-human," preferring instead to think of human-environment interactions in terms of the more organic "composting"; see esp. *Staying With the Trouble*, 102.
[38] Sheryl Vint, *Animal Alterity: Science Fiction and the Question of the Animal* (Liverpool: Liverpool University Press, 2010). Vint renders the breakdown of the human-animal dichotomy in works of speculative fiction (in which one often finds apocalypses), providing another rich avenue of inquiry.
[39] Morton, Humankind, 133.
[40] Zakiyyah Iman Jackson, *Becoming Human: Matter and Meaning in an Antiblack World* (New York: New York University Press, 2020), Introduction, Apple Books.
[41] Ward, *Salvage the Bones,* chap. 4.

dogs to increase his family's social and economic status.[42] Yet the relationship between Skeetah and China indicates a mutual understanding of their intertwined, precarious existence. This situates the novel within a discourse of "becoming human" in which it

> intervenes productively in reconsidering the role of "the animal" or the "animalistic" in the construction of "the human" by producing nonbinary models of human-animal relations, advancing theories of trans-species interdependency, observing trans-species precarity, and hypothesizing cross-species relationality in a manner that preserves alterity while undermining the nonhuman and animality's abjection, an abjection that constantly rebounds on the marginalized humans.[43]

The cross-species relationality of Skeetah and China throughout the novel contrasts sharply for its problematic operation within a discourse of capital and bare life. At its core sit a number of hard decisions that nevertheless produce agency through the ability to respond—or responsibility—on own conditions. China decides which of her puppies to nurture and on the terms of this nurturing, with persuasion by Skeetah,[44] and she is consulted before Skeetah agrees to fight her. In a moment of consultation "unhooks the dull heavy chain from China's neck, smiles; she smiles with him."[45] The human-animal binary evaporates in a moment of emancipation and communal determination. This indicates an agential response to being treated as an object of production, which Danowski and Viveiros de Castro argue is necessary to produce a radically different configuration of human and nonhuman subjects (and capital): "the only way to conjure an Outside is to produce it from the inside by driving the capitalist machine into overdrive, to accelerate the acceleration that defines it, to maximize the creative destruction that moves it until it finishes by destroying itself and creating for us a radically *new* world."[46] China fights to the point of nearly killing herself as she refuses reduction and insists on establishing her subjectivity. Deciding when to kill and when to die affects her motherhood, as well. During the hurricane, China leaves her puppies and the family to leap into the rising floodwaters in an act of apparent combined suicide and abandonment. It could be asserted that Ward is

---

[42] Ibid., chap. 8.
[43] Jackson, *Becoming Human*, Introduction.
[44] Ward, *Salvage the Bones*, chap. 6.
[45] Ibid., chap. 8.
[46] Déborah Danowski and Eduardo Viveiros de Castro, *The Ends of the World*. Trans. Rodrigo Nunes (Cambridge: Polity, 2017), 51-52 (original emphasis).

echoing the precarious position of Black Americans in the rural South in the way that she positions the family's companion animal in *Bones*. Both are dehumanized from a dominant cultural perspective of history in such a reading, and as such, it would be impossible not to think here of Margaret Garner's act of infanticide as saving grace, popularized in Toni Morrison's *Beloved*. In that novel, a self-emancipated Black mother encounters patrollers who come to re-enslave her family. Instead of submitting, she kills one child and attempts to do the same to the others. Morrison makes clear her position that she believes Garner to have acted in good faith in an effort to prevent her family from "the living death that was slavery," even if this meant ascribing their future to a projected afterlife.[47] Killing to avoid being killed (or killing one's offspring to avoid their being killed) constitutes an important restoration of agency for those dehumanized in American history, as Edwidge Danticat notes in *The Art of Death: Writing the Final Story*. "Freedom, like death," she writes, "can be defined in different ways."[48] There exists an agency in choosing how to die (or how to live). On some level, choosing death constitutes a refusal to accept the end without dictating at least some of the terms of its execution. As Danticat earlier notes, it is an historic resituating of life and death on one's own terms, which restores (at least some of the) agency lost in the dehumanizing act of enchattlement.[49] As such, even a pessimistic reading of China—and *Bones* more broadly—that equates animal life and Black life with bare life does so not to maintain a hierarchy of humanity, but to acknowledge a plurality of subject positions and claims on the agency to live and die. This situates the novel within the framework of the posthuman turn and acknowledges the present postapocalyptic lives that result from the radical transformation of the apocalypse and its aftermath.

Conclusion

Millennial apocalypse and climate catastrophe combine in *Salvage the Bones* to expand existing narrative structures and the human beings they center. In an interview with the *Paris Review* Jesmyn Ward explains that "salvage" in her title conveys a sense of claiming whatever remains in the aftermath of environmental (world-)destruction: "there is honor in that term. It says that come hell or high water […] you are strong, you are fierce, and

---

[47] Qtd. in Edwidge Danticat, *The Art of Death: Writing the Final Story* (Minneapolis: Graywolf Press, 2017), 105.
[48] Danticat, *The Art of Death,* 115-16.
[49] Ibid., 79.

you possess hope. When you stand on a beach after a hurricane [...] and all you have are your hands, your feet, your head, and your resolve to fight, you do the only thing you can: you survive. You are a savage."[50] The connection between savaging humanity and salvaging humanity implicates what I have described as a posthuman turn toward radically reconfigured subjectivities in *Salvage the Bones*. This shifts the millennialist apocalypse tradition significantly without completely evaporating the biblical influence, which Ward insists remains an important influence in her work.[51] The aftermath of the storm in *Bones* speaks to what Berger identifies in Revelation 21:15-17 as the inability to measure the course of change in the altered world "with human measurements." A "discourse entirely of the other" cannot be measured, he argues; "we should rather say that the jagged discursive relations are conflicts, struggles, and misunderstandings—not that they are instances of incommensurability."[52] I would add that the privileged position of the prophetic writer (and often the reader, as well) of *Revelation* fails to admit the "entirely other" of its discursive mode because he has not considered a space in which to do so. An expanded reading of millennial (post)apocalypse and its posthuman beings would be needed, and Ward provides this. Such a reading includes the pre-(which is also always post-)apocalyptic anxiety at the family's pit bull birthing puppies. "I never thought I'd get five, Esch. With it being her first, I thought I'd get two, maybe. I figured she trample them or that they'd just come out dead. But I never thought she'd let me save so many."[53] Read elsewhere in the novel as a fierce warrior and frequent victor in the dog fights of the Swamp Pit, here China proves the incommensurability of the nonhuman other. Skeetah initially struggles to reconcile China's transformation because of his (the family's) reliance on her, in part, as an object of capital. Yet the novel also expands their relationship beyond economic salvage. The imagery of China as a messianic figure who comes to save the family despite being consistently sacrificed to the apocalypse leaves Skeetah similarly hollowed out, a posttraumatic shell of his former self.

The novel concludes with another messianic image, that of China returning, illuminated in a way reminiscent of the mythic glory of the ancient heroines Esch so admires. This reads not as a message of hope, but as a reminder that resistance, destruction and renewal are part of life for

---

[50] Jesmyn Ward, "Jesmyn Ward on Salvage the Bones," *Paris Review*, August 30, 2011.
[51] Ibid.
[52] Berger, *After the End*, 13.
[53] Ward, *Salvage the Bones*, chap. 2.

postapocalyptic worlds and the posthuman beings who inhabit them.[54] No future rapture will save the unchosen; they know only the apocalypse, and birth new—impossibly unimaginable, yet nevertheless real—worlds. Esch and the "smaller animals" are admittedly not the "other animals"—the people chosen to inherit a millennial kingdom and escape violence—nor are they nonbelievers. Nor are they supreme subjects of enlightenment humanism revealed to be at the end of history. Through the experience of living through repeated (thus post)apocalypse they reconfigure millennialist classification altogether. Ward exposes radically new subjectivities that extend beyond (and exist apart from) traditional concepts of the human. The postapocalyptic lives in *Bones* suggest Braidotti's critical posthuman subject. They echo the extended subjectivity of "humankind" for which Morton advocates. And they indicate the modes of "becoming human" that Jackson situates at the center of a fuller recognition of Black (and animal) humanity. This reconfigures millennialist thinking about the end of the world in meaningful ways: it disconnects eschatological thinking from purely futuristic projections and resituates apocalypse in an enduring present; and it reconsiders a failed concept of "chosen people" by instead centering those unchosen who nevertheless work through the destruction. Ward embeds new meaning in her rendering of the post-apocalypse and its determinations for the unchosen of Enlightenment thinking, even as she performs the work of establishing new ways of thinking about how narratives (re-)present and (re-)produce the conceptual frameworks with which they engage. *Bones* rethinks the paradigm of postapocalyptic anxiety to realize reconfigured modes of being in the world that consider human and other-than-human agents without the baggage of human / nature opposition. Here Ward presents a new world that re-members the old one, still looking to a renewed future with anxious hope that the next apocalypse will not be the end of the world.

References

Aune, David. "Revelation: Introduction." In *The Harper Collins Study Bible: New Revised Standard Version*, edited by Wayne A. Meeks, 2307-09. (New York: Harper Collins, 1993).
Bercovitch, Sacvan. [1978] 2011. *The American Jeremiad*. Madison: University of Wisconsin Press.
—. [1975] 2011. *Puritan Origins of the American Self*. New Haven: Yale University Press.

---

[54] Cf. Ward, "In Salvage The Bones, Family's Story Of Survival."

Berger, James. 1999. *After the End: Representations of Post-apocalypse.* Minneapolis: University of Minnesota Press.
Braidotti, Rosi. 2013. *The Posthuman.* Cambridge: Polity.
Danowski, Déborah, and Viveiros de Castro, Eduardo. *The Ends of the World.* Translated by Rodrigo Nunes. Cambridge: Polity, 2017.
Danticat, Edwidge. 2017. *The Art of Death: Writing the Final Story.* Minneapolis: Graywolf Press.
Encyclopædia Britannica. "Jim Jones." Accessed March 23, 2021. https://www.britannica.com/biography/Jim-Jones.
Fukuyama, Francis. 1992. *The End of History and the Last Man.* New York: The Free Press.
Gilroy, Paul. 2000. *Against Race: Imaging Political Culture Beyond the Colour Line.* Cambridge: Harvard University Press.
Haraway, Donna. 2016. *Manifestly Haraway.* Minneapolis: University of Minnesota Press.
—. 2017. *Staying with the Trouble: Making Kin in the Chthulucene.* Durham: Duke University Press.
The Harper Collins Study Bible: New Revised Standard Version. New York: Harper Collins, 1993.
Jackson, Zakiyyah Iman. *Becoming Human: Matter and Meaning in an Antiblack World.* New York: New York University Press, 2020. Apple Books.
Morrison, Toni. [1987] 2005. *Beloved.* London: Vintage.
Morton, Timothy. 2017. *Humankind: Solidarity with Nonhuman People.* London: Verso.
Strozier, Charles. 1994. *Apocalypse: on the Psychology of Fundamentalism in America.* Boston: Beacon Press.
Vint, Sheryl. 2010. *Animal Alterity: Science Fiction and the Question of the Animal.* Liverpool: Liverpool University Press.
Ward, Jesmyn. "Jesmyn Ward on *Salvage the Bones.*" Interview by Elizabeth Hoover. *Paris Review*, August 30, 2011. https://www.theparisreview.org/blog/2011/08/30/jesmyn-ward-on-salvage-the-bones/.
—. "In *Salvage The Bones*, Family's Story of Survival." Interview by Michel Martin. *Tell Me More: Books*, NPR, December 5, 2011. http://www.npr.org/templates/story/story.php?storyId=143141874&t=1559551202406.
—. *Salvage the Bones.* New York and London: Bloomsbury, [2011] 2017. Apple Books.

Wilson, Woodrow. [1902] 2001. "The Ideals of America." In *The American Intellectual Tradition*, vol. II, fourth ed., edited by David A. Hollinger and Charles Capper, 123-30. Oxford: Oxford University Press.

Wojcik, Daniel. 1997. *The End of the World as We Know It: Faith, Fatalism, and Apocalypse in America*. New York: New York University Press.

Wolfe, Cary. 2010. *What is Posthumanism?* Minneapolis: University of Minnesota Press.

Chapter Thirteen

The Post-Apocalyptic Aesthetics of Emily St. John Mandel's *Station Eleven*

Murat Kabak

Aesthetics and politics stand in opposition to one another from the origin of the aesthetics in the eighteenth century to the literary and philosophical correspondences in the first half of the twentieth century. The Egyptian-Greek poet Constantine P. Cavafy illustrates this opposition between these contesting realms in a poem titled, "Dareios", where the fictional poet Phernazis writes Greek poetry in the middle of a war with the Romans.[1] The question of whether one should go on writing poetry in times of crisis is relevant to a discussion of artistic creation and appreciation after the apocalypse. While the poetics of various genres and art forms have been widely discussed in literary studies, the study of aesthetic creation has been ignored in studies in post-apocalyptic fiction. The main reason behind this lack of post-apocalyptic poetics is the overt political associations of this genre's form and content. In her study of postcolonial poetics, for instance, Elleke Boehmer refers to the same problem: How could "writing be both political and concerned with formal and even aesthetic principles?"[2] In other words, when there is a more immediate political goal or an emancipatory agenda, can we talk about poetics? By adapting the same formulation to post-apocalyptic fiction, we may ask: when human beings are reduced to a state of survival, can we talk about beauty? After the catastrophe, what happens to the artwork that aims to serve no function but to be itself? Or if

---

[1] Stathis Gourgouris, "*Poiein* — Political Infinitive," *PMLA/Publications of the Modern Language Association of America* 123, No. 1 (2008): 225-227.
[2] Boehmer, Elleke. "Question of Postcolonial Poetics". In *Postcolonial Poetics: 21st-Century Critical Readings*, (Basingstoke: Palgrave Macmillan, 2018), 19.

we put *Station Eleven*'s oft-repeated motto into question form: If survival is insufficient, how do we position beauty in the context of the apocalypse?

Emily St. John Mandel's post-apocalyptic pandemic novel, *Station Eleven* (2014), is currently enjoying a surge of popularity after the coronavirus pandemic. Even Mandel expressed her disbelief in a recent podcast interview by playfully asking: "Why would anybody in their right mind want to read *Station Eleven* during a pandemic?"[3] The rising interest, however, is evident in the apocalyptic genre fiction from Cormac McCarthy's *The Road* (2006) to Josh Malerman's *Bird Box* (2014), or their coming-of-age variants such as Suzanne Collins's *The Hunger Games* trilogy (2008-2010) that are often mass-produced and adapted to film and television. As opposed to many previous representatives of the post-apocalyptic genre, some critics champion Mandel's novel *Station Eleven* for its hopeful vision of humanity because the novel avoids nearly all conventions of the genre: "ragged bands of survivors; demolished urban environments surrounded by depleted countryside; defunct technologies; desperate scavenging; poignant yearning for a lost civilization, often signified by the written word; and extreme violence, including cannibalism, enacted by roving gangs of outlaws."[4] With the absence of such survivalist motifs, some critics argue that the novel emphasizes "the importance and value of art and memory."[5] In the present chapter, I aim to analyze the political implications of this hopeful vision, and the emphasis on arts in Mandel's *Station Eleven*. I argue that the novel offers an ambiguous case through its rejection of the understanding of a unique work of art that is to be contemplated for its own sake. Instead, it returns to the notion of a ritualistic artwork in the context of its relations to its audience through emphasizing the therapeutic potential of art. Yet, this positive note on arts does not necessarily involve a critique of modernity or some sort of societal change. While the novel invites the reader to reevaluate the position and agency of art, it also remains complicit in modernity due to its nostalgia for the modern world.

---

[3] Emily St. John Mandel, "Emily St. John Mandel on Fact, Fiction, and the Familiar (Episode 92)," interview by Tyler Cowen, April 8, 2020, https://medium.com/conversations-with-tyler/emily-st-john-mandel-tyler-cowen-pandemic-covid-19-fiction-author-703c65c12447.

[4] Hicks, Heather. J. "Introduction: Modernity Beyond Salvage". In *The Post-Apocalyptic Novel in the Twenty-First Century: Modernity Beyond Salvage*, (New York: Palgrave Macmillan, 2016), 6.

[5] Maximilian Feldner, "'Survival is Insufficient': The Postapocalyptic Imagination of Emily St. John Mandel's *Station Eleven*," *Anglica: An International Journal of English Studies* 27, No. 1 (2018): 166.

"Beauty in the altered world": Shakespeare and the Traveling Symphony

Station Eleven opens with a Canadian production of *King Lear* on the day when the so-called "Georgia Flu" hits North America and the novel jumps back and forth between two timelines: before the virus outbreak and twenty years after the pandemic devastates 99.99 percent of the world population. What is left of civilization is "an archipelago of small towns" whose inhabitants "had fought off ferals, buried their neighbors, lived and died and suffered together in the blood-drenched years ... and then only by holding together into the calm."[6] Interestingly, this is the only reference to the atrocities that were committed after the collapse of civilization in the novel. Instead, the post-collapse section of the novel is centered around the journey of a traveling troupe called the Traveling Symphony performing a selection of Shakespeare plays around the Lakes Michigan and Huron. In that sense, when compared to the other representatives of the post-apocalyptic genre fiction, *Station Eleven* is uncharacteristically silent about the causes of the plague. For instance, in Margaret Atwood's *MaddAddam* trilogy (2003-2013), which has been hailed by the critics as a prescient critique of human-centered thought, the JUVE is a genetically engineered virus. Or more recently, another Canadian author, Saleema Nawaz deals with the public reactions to the virus outbreak, and the ineffective government responses to the pandemic in *Songs for the End of the World* (2020). Mandel's *Station Eleven*, however, only alludes to the cataclysmic events in a passing reference to people who "survived against unspeakable odds."[7] The novel foregrounds the regenerative aspect of arts after the cataclysm but the intertextual connections to Shakespeare are problematic in terms of the suggestions of the durability of artwork after the apocalypse.

The sections of the novel that deal with an aesthetic community after the apocalypse employ Shakespeare's plays as a conceptual framework. As the first chapter ends with the death of the actor who played the king, the second chapter, titled "A Midsummer Night's Dream", opens with the post-collapse theatre troupe preparing a scene for Shakespeare's famous comedy. Philip Smith observes that if *"King Lear* heralds the apocalypse, then *A Midsummer Night's Dream* heralds the possibility of rebirth."[8] However, while it is thematically significant, this possibility suggests a very Eurocentric rebirth. In several instances, the novel draws attention to the theme of plague

---

[6] Emily St. John Mandel, *Station Eleven* (New York: Vintage Books, 2015), 48.
[7] Ibid, 48.
[8] Philip Smith, "Shakespeare, Survival, and the Seeds of Civilization in Emily St. John Mandel's *Station Eleven,*" *Extrapolation* 57, No. 3 (2016): 294.

in Shakespeare's oeuvre, especially in the staging of the second act of *Midsummer*:

> What was lost in the collapse: almost everything, almost everyone, but there is still such beauty. Twilight in the altered world, a performance of *A Midsummer Night's Dream* in a parking lot …
> 'Therefore the winds, piping to us in vain, as in revenge, have sucked up from the sea contagious fogs…'
> *Pestilential*, a note … explains, next to the word *contagious*, in … the text that the Symphony carries. Shakespeare was the third born to his parents, but the first to survive infancy. Four of his siblings died young. His son, Hamnet, died at eleven and left behind a twin. Plague closed the theaters again and again, death flickering over the landscape. And now in a twilight once more lit by candles, the age of electricity having come and gone, Titania turns to face her fairy king …
> 'And through this distemperature, we see the seasons alter.'[9]

The third-person narrator's description of the loss is modified with the adverb "almost" because what remains is the beauty of the performance in "the altered world". The striking association in the passage is the company's association with Shakespeare the survivor who had been haunted by the plague for all his life but his work was able to transcend yet another plague in the twenty-first century. As the final line of the chapter suggests, "survival is insufficient,"[10] and this lack has been fulfilled by the revival of the Shakespeare performance. The way the novel enacts the play, however, suggests more than a hope of rebirth.

While there are no other temporal clues in the passage, the repetition of "twilight" suggests that the play is being staged in the evening twilight. This brief period between sunset and dusk indicates that the sublime performance provides a temporary respite, but there is an even darker and more serious threat than the pandemic. This threat is the presence of a cult leader who chases down the troupe because one of his child brides has run away with the Traveling Symphony without their knowledge. Not only with the repetition of plague symbolism but with the Titania-Oberon dispute over the changeling, the novel consolidates its connection to Shakespeare. The novel is to resolve the tension between the company and the prophet just as the play resolves the discordance in nature by Oberon seizing the changeling. While the prophet's presence disrupts their performance, his death at the end allows the company "[to] perform Shakespeare and

---

[9] Mandel, *Station Eleven*, 57-58.
[10] Ibid, 58.

music"[11] once again. This reconciliation, however, simplifies the initial conflict of Shakespeare's *A Midsummer Night's Dream* which is problematic for two reasons. First, the probable causes of the pandemic have been obscured. While the passage above quotes Titania's lines in Act 2, Scene 1, lines 90 to 110, the Queen's final lines in her address to Oberon are omitted: "And this same progeny of evils comes / From our debate, from our dissension; / We are their parents and original."[12] Here, even the Queen of the Faeries accepts her part in the dispute over the empire's colonial dream, i.e., the possession of the Indian changeling. Whereas the novel does not even acknowledge the role of the human involvement in ecological imbalance, such as the sea of contagious fogs, distemperature (the pandemic), or the seasons' alteration (climate change). Secondly, this dismissal of the human factor reduces the conflict at the center of the novel to a cause between the defenders of arts and culture (the Traveling Symphony) and the uncivilized, barbaric religious cult of the prophet.

An even more crucial implication of such a reduction in the context of civilization versus barbarity is the choice of its main influence. Samuel Johnson famously claims that even though Shakespeare's "personal allusions, local customs, or temporary opinions, have … been lost"[13] to his present age, Shakespeare's plays tap into the core of humanity because:

> His characters are not modified by the customs of particular places, unpractised by the rest of the world … or by the accidents of transient fashions or temporary opinions: they are the genuine progeny of common humanity, such as the world will always supply; and observation will always find.[14]

By placing Shakespeare as a playwright who can reach an audience even twenty years after a global pandemic that devastates 99. percent of the world population, the novel echoes Johnson's Enlightenment universalism: "[The troupe] performed more modern plays sometimes in the first few years, but what was startling … was that audiences seemed to prefer Shakespeare to their other theatrical offerings … 'People want what was best about the world.'"[15] While other forms of art also survive, the durability

---

[11] Ibid, 331.
[12] William Shakespeare, *A Midsummer Night's Dream* (New York: Simon & Schuster Paperbacks, 2009), 2.1.118-120.
[13] Johnson, Samuel. "Preface to Shakespeare". In *The Norton Anthology of Theory and Criticism,* ed. Vincent. B. Leitch (New York: W. W. Norton & Company, 2001), 469.
[14] Ibid, 469.
[15] Mandel, *Station Eleven*, 38.

of Shakespeare is prioritized, and the survival of his works do not only suggest the longevity of his works "but the survival of all anglophone culture through Shakespeare."[16] Exposure to Shakespeare, at the end of the narrative, inevitably brings civilization to the wasteland of the novel. By rendering the Bard as its primary influence, the novel suggests that arts and culture may act as a remedy against the destructive consequences of the plague. However, this presupposition, as I have illustrated, does not only avoid the human factor in the apocalypse but is also universalist and Western-oriented. The novel shifts its attention from the work of art as a timeless and universal artefact, to objects without any particular use-value after the apocalypse.

The work of art in the age of post-apocalypse

While Shakespeare survives the apocalypse, some products of modernity are lost in the fictional world of *Station Eleven*. In the concluding section of the first chapter, the third-person narrator provides a list of missing items after the pandemic that serves as a bridge between the pre- and post-pandemic worlds of the novel:

> No more trains running under the surface of cities on the dazzling power of the electric third rail. No more cities. No more films ... No more pharmaceuticals. No more certainty of surviving a scratch on one's hand ... No more flight. No more towns glimpsed from the sky through airplane windows, points of glimmering light; no more looking down from thirty thousand feet and imagining the lives lit up by those lights at that moment ... No more Internet. No more social media.[17]

As the point of view changes from under the surface of the earth to the skies, the narrator provides an account of how the human relationship with distances was altered by the pandemic both at macro and micro levels. The tone is nostalgic rather than critical towards such products of modernity. We may compare this to a similar instance in Margaret Atwood's *Oryx and Crake* (2003) where the protagonist, Jimmy, recalls his childhood in the pharmaceutical compounds where the adults were yearning for their lives before the ecological collapse: "*Remember when you could drive anywhere? Remember when everyone lived in the pleeblands? Remember when you*

---

[16] Smith, "Shakespeare, Survival, and the Seeds of Civilization in Emily St. John Mandel's *Station Eleven*," 298.
[17] Mandel, *Station Eleven*, 31-32.

could fly anywhere in the world, without fear? Remember hamburger chains, always real beef, remember hot-dog stands? ... Remember when voting mattered?"[18] The repetition of "remember" in the passage reveals the insistence on having a sense of control over one's life and memories because these questions are posed rhetorically. The interlocutors who exchange these questions do not expect an answer to the inquiry. Instead, they are expressing their discontent with their current lives. However, throughout Atwood's trilogy, this nostalgia for the pre-collapse world of modernity is satirized by pointing out that this insistence on human-centered thought ultimately led to ecological collapse. In Mandel, however, there is no reference or indication that the intercontinental travel and communication, the excess of the human carbon footprint caused environmental degradation or at least accelerated the transmission of the virus. Not only ecological consciousness but also any critique of modernity is absent. Instead, the power of modernity to bring distances into proximity is celebrated. This celebration of human enterprise and the longing for its products reach its height in the chapters concerning a "Museum of Civilization" that is born out of the ruins of the old world.

Station Eleven is invested in the beauty of everyday objects from a paperweight to a snow globe. One of the key chapters in the novel involves a group of survivors who are stranded in an airport. Completely abandoned at the airport with no contact from the outside, their hope of the arrival of a rescue team or any communication grows thin. Separated from the cataclysm outside, the airport survivors build an alternative community with its laws and regulations. Clark Thompson, one of the survivors who is a middle-aged lawyer, decides to pile all the "useless objects" on a counter at the airport:

> Clark placed his useless iPhone on the top shelf ... [and] added his laptop, and this was the beginning of the Museum of Civilization. He mentioned it to no one, but when he came back a few hours later, someone had added another iPhone, a pair of five-inch red stiletto heels, and a snow globe.
>
> Clark had always been fond of beautiful objects, and in his present state of mind, all objects were beautiful. He stood by the case and found himself moved by every object he saw there, by the human enterprise each object had required. Consider the snow globe. Consider the mind that invented those miniature storms, the factory

---

[18] Margaret Atwood, *Oryx and Crake* (New York: Nan A. Talese, 2003), 63 (emphasis in original).

worker who turned sheets of plastic into white flakes of snow, the hand that drew the plan for the miniature Severn City.[19]

With the absence of electricity and mass communication, neither smartphones nor laptops have any practical use in the novel. This impromptu museum is what Michel Foucault calls heterotopia in his article "Of Other Spaces". Tracing the development of the museum in the nineteenth century, Foucault argues that the idea of systematizing not only knowledge but of "all times, all epochs, all forms, all tastes, [and] the idea of constituting a place of all times that is itself outside of time and inaccessible to its ravages"[20] is an acutely modern concept. From technological gadgets to stiletto heels, these objects that are useless after the apocalypse are brought together in an attempt at commemorating the greatest human achievements by preserving them from the ravages of time. Indeed, this silent agreement between the survivors when they contribute to the museum's inception is indicative of the reverence paid to these objects. A little earlier in the novel, we have another scene where Thompson exalts all communication and transportation technology into the state of a miracle: "This was how [Clark] arrived in this airport: he'd boarded a machine that transported him at high speed a mile above the surface of the earth ... These taken-for-granted miracles that had persisted all around them."[21] This religious attribution to *technē* should remind the reader of the miracles that pervade their mundane everyday life.

Another way of thinking about this numinous sense surrounding the works of art in the novel is Walter Benjamin's concept of aura because Thompson's celebration of human enterprise complicates the reception of the artwork "in the age of mechanical reproduction". In his famous essay, Benjamin traces the concept of aura in three stages.[22] In the first stage, the work of art has ritual value, and its aura or its "distance, however close it may be"[23] is still intact. The second stage or the "secular cult of beauty"[24] sets the autonomous work of art from its ritual function. For Benjamin, the object's aura is completely diminished by the third stage of mechanical reproducibility where there is neither authenticity nor any form of distance

---

[19] Mandel, *Station Eleven*, 254-255.
[20] Michel Foucault and Jay Miskowiec, "Of Other Spaces," *Diacritics* 16, no. 1 (1986): 26.
[21] Mandel, *Station Eleven*, 232-233.
[22] Benjamin, Walter. "The Work of Art in the Age of Mechanical Reproduction". In *Illuminations*, ed. Hannah Arendt (New York: Schocken, 1969), 223-225.
[23] Ibid, 222.
[24] Ibid, 223.

is possible.[25] However, *Station Eleven* complicates Benjamin's account on two grounds because the artworks in the novel still have a ritualistic value. Firstly, it is important to note that Thompson's appraisal does not involve a "disinterested love of a free play of the mind on all subjects, for its own sake"[26] and the snow globe is not an art form without intent, as it was assumed in the second stage of autonomous artwork. While the airport inhabitants place these objects into a museum space, their avoidance of positing the objects as unique works of art to be contemplated for their own sake suggests that these materials have a more immediate instrumental purpose. Going back to the initial quote, what Thompson praises about the snow globe is less to do with the formal characteristics of the globe and more to do with the genius who created it and the traces of human labor invested in its production. Secondly, even though the laptops, smartphones, or more specifically, the snow globe, are the products of mechanical reproduction, hence, are supposed to be stripped of their aura, they serve a ritualistic role. As suggested in his deliberation "over the mind that invented,"[27] these objects, from the curator's point of view, celebrate human achievement and intelligence. The personal and subjective nature of the artwork is attributed to the body of people who produced it. In that sense, the snow globe on the counter does not refer to itself but stands for the accumulation of human knowledge and connections.

In his reply to Benjamin's "Mechanical Reproduction", Theodor Adorno proposes an alternate model of aura in his *Ästhetische Theorie* (1970). As opposed to Benjamin's emphasis on the objects of technology, Adorno considers other non-human objects as auratic as well, and the possibility of encountering the non-human object's aura depends on a reciprocal activity. For Adorno, the "experience of aura rests on the transposition of a response … To perceive the aura of an object means to invest it with the ability to look at us in return."[28] Thompson's engagement with the aesthetic object, however, does not allow the object to look back because he is "moved … by the human enterprise each object had required". In that sense, the engagement with the work of art is only instrumental in celebrating human genius.

---

[25] Ibid, 233.
[26] Arnold, Matthew. "The Function of Criticism at the Present Time". In *The Norton Anthology of Theory and Criticism*, ed. Vincent. B. Leitch (New York: W. W. Norton & Company, 2001), 814.
[27] Mandel, *Station Eleven*, 255.
[28] Yvonne Sherratt, "Adorno's Aesthetic Concept of Aura," *Philosophy & Social Criticism* 33, No. 2 (2007): 170.

Conclusion: "The thought of ships moving over the water"

The novel moves from the theatrical community around Shakespeare to beautiful but useless objects in its exploration of what is human. When stripped of all their technology, civilization, and industry, the post-collapse human communities are forced to survive. The avoidance of survivalist themes and motifs is an aesthetic choice to emphasize the role of arts and culture in patching up humanity's wounds twenty years after a global catastrophe. The theatrical company, however, is threatened by a disruption of the shared experience. In that sense, the conflict between the Traveling Symphony and the prophet's cult formulates a simple dichotomy of arts and culture versus barbarity, which is resolved at the end by the elimination of evil. Shakespeare, not only surviving the plague in his time, survives the fictional pandemic in the novel as well. However, promising a Eurocentric vision of rebirth, this survival does not offer a multicultural and diverse future.

Closely connected to this modernist view, the novel is also nostalgic for the products, industry, and technology of the modern world to the extent that the objects of interest are exalted into ritualistic objects to celebrate their human elements. A new form of aesthetics emerges out of the ruins of the old world, but this new aesthetics diverges from both Benjamin and Adorno's accounts of aura. The laptops with dead batteries, paperweights, and snow globes are not beautiful due to a reciprocal relationship between the spectator and the object. Even though they were mechanically reproduced, and brought closer to their spectators, their aura in the novel survives both the apocalypse and their mechanical reproducibility. Instead, the novel suggests that their aura ultimately emanates from their human spectators who see in them the traces of human labor.

The novel's concluding paragraph provides yet another hopeful but ambiguous vision. Clark Thompson the curator sits in his chair in the museum and reflects on the future of humanity:

> He has no expectation of seeing an airplane rise again in his lifetime, but is it possible that somewhere there are ships setting out? ... Perhaps vessels are setting out even now, traveling toward or away from him, steered by sailors armed with maps and knowledge of the stars, driven by need or perhaps simply by curiosity ... If nothing else, it's pleasant to consider the possibility. He likes the thought of ships moving over the water, toward another world just out of sight.[29]

---

[29] Mandel, *Station Eleven*, 332-333.

Smith reads the ending as suggestive of the novel's colonial aspirations due to extended references to Shakespeare [30], as the sailors are "armed with maps and knowledge". These ships, however, might as well be referring to an earlier mention of a "fleet lay at anchor off the coast of Malaysia"[31] due to the 2008 economic collapse. When the economic crisis hits the intercontinental trade, these ships that are temporarily suspended off the coast are left with only "a skeletal crew walking the empty rooms and corridors."[32] In that sense, Thompson's reimagining of the empty cargo freights as vessels full of curious sailors is yet another rebirth of the modern market economy.

In its exploration of the work of art in a post-catastrophe civilization, Emily St. John Mandel's *Station Eleven* presents a curious case. Instead of indulging in depictions of a fictional pandemic that devastates most of the world's population, the novel foregrounds the regenerative aspect of arts after the cataclysm. However, the intertextual connections to Shakespeare are problematic in terms of the suggestions of the durability of the artwork after the apocalypse. The survival of the exclusively Western cultural artifacts suggests a yearning for the products of modernity. As the central concern of climate fiction, climate change "is not shaped or reflected by literary form, but emerges through the interaction of literary and non-literary objects."[33] In *Station Eleven*, however, this interaction exposes the novel's particularly Western and modern notions of technology, arts, and culture. As suggested earlier, this interaction also symbolically revives the engine of the capitalistic drive. Rather than freeing arts, culture, and civilization from strict aesthetic principles, *Station Eleven* instrumentalizes them as a means to champion human enterprise.

References

Arnold, Matthew. "The Function of Criticism at the Present Time". In *The Norton Anthology of Theory and Criticism,* edited by Vincent. B. Leitch, 806-825. (New York: W. W. Norton & Company, 2001).

Atwood, Margaret. 2003. *Oryx and Crake.* New York: Nan A. Talese.

---

[30] Smith, "Shakespeare, Survival, and the Seeds of Civilization in Emily St. John Mandel's *Station Eleven*," 301.
[31] Mandel, *Station Eleven*, 28.
[32] Ibid, 28.
[33] Pieter Vermeulen, "Beauty That Must Die: *Station Eleven*, Climate Change Fiction, and the Life of Form." *Studies in the Novel* 50, no. 1 (2018): 9.

Benjamin, Walter. "The Work of Art in the Age of Mechanical Reproduction". In *Illuminations*, edited by Hannah Arendt, 217-251. Translated by Harry Zohn. (New York: Schocken, 1969).

Boehmer, Elleke. "Question of Postcolonial Poetics". In *Postcolonial Poetics: 21st-Century Critical Readings*, 19-38. (Basingstoke: Palgrave Macmillan, 2018).

Feldner, Maximilian. "'Survival is Insufficient': The Postapocalyptic Imagination of Emily St. John Mandel's *Station Eleven*." *Anglica: An International Journal of English Studies* 27, no. 1 (2018): 165-179. https://Doi.org/10.7311/0860-5734.27.1.12 2018

Foucault, Michel, and Jay Miskowiec. "Of Other Spaces." *Diacritics* 16, no. 1 (1986): 22-27. https://doi.org/10.2307/464648

Gourgouris, Stathis. "*Poiein* — Political Infinitive." *PMLA/Publications of the Modern Language Association of America* 123, no. 1 (2008): 223-228. https://doi.org/10.1632/pmla.2008.123.1.223

Hicks, Heather. J. "Introduction: Modernity Beyond Salvage". In *The Post-Apocalyptic Novel in the Twenty-First Century: Modernity Beyond Salvage*, 1-26. (New York: Palgrave Macmillan, 2016).

Johnson, Samuel. "Preface to Shakespeare". In *The Norton Anthology of Theory and Criticism,* edited by Vincent. B. Leitch, 468-480. (New York: W. W. Norton & Company, 2001).

Mandel, Emily St. John. "Emily St. John Mandel on Fact, Fiction, and the Familiar (Episode 92)." Interview by Tyler Cowen, April 8, 2020, https://medium.com/conversations-with-tyler/emily-st-john-mandel-tyler-cowen-pandemic-covid-19-fiction-author-703c65c12447

Mandel, Emily St. John. 2015. *Station Eleven.* New York: Vintage Books.

Shakespeare, William. 2009. *A Midsummer Night's Dream.* New York: Simon & Schuster Paperbacks.

Sherratt, Yvonne. "Adorno's Aesthetic Concept of Aura." *Philosophy & Social Criticism* 33, no. 2 (2007): 155-177. https://doi.org/10.1177/0191453707074137

Smith, Philip. "Shakespeare, Survival, and the Seeds of Civilization in Emily St. John Mandel's *Station Eleven*." *Extrapolation* 57, no. 3 (2016): 289-303. https://doi.org/10.3828/extr.2016.16

Vermeulen, Pieter. "Beauty That Must Die: *Station Eleven*, Climate Change Fiction, and the Life of Form." *Studies in the Novel* 50, no. 1 (2018): 9-25. https://doi.org/10.1353/sdn.2018.0001.

CHAPTER FOURTEEN

FAILING ECOSYSTEMS, CONTAGION, AND CATASTROPHE IN ZADIE SMITH'S 'NARRATIVES OF ENTROPY': *THE CANKER* AND *ELEGY FOR A COUNTRY'S SEASONS*

MANASVINI RAI

Introduction

Deconstructing environmental crises in the Anthropocene epoch and an imaginary mise-en-scène, the transatlantic black British author Zadie Smith explores through an intermingling of the poetic, the political, and the universal, the causality associated with devolving ecosystems at the brink of catastrophe. Smith's allegorical short story *The Canker* (2019) and her essay *Elegy for a Country's Seasons* (2018) that are examined in this study may be seen as forms of the "dystopian imaginary,"[1]—a concept defined by Kurasawa as being equivalent to Susan Sontag's phrase "the imagination of disaster."[2] These imaginaries present a potential for foresight and are concerned with the depiction of varied futures for humankind and the planet. These futures involve debatable probabilities pertaining to outcomes characterized either by balance or chaos in biological and social ecosystems. Scenarios envisioned by such apocalypse narratives

---

[1] Fuyuki Kurasawa, "Cautionary Tales: The Global Culture of Prevention and the Work of Foresight," *Constellations* 11, no. 4 (2004): 458, http://www.yorku.ca/kurasawa/Kurasawa%20Articles/Constellations%20Article.pdf.

[2] Susan Sontag, "The Imagination of Disaster, " in *Against Interpretation and Other Essays* (New York: Dell Publishing, 1966), 217, 226, quoted in Kurasawa, "Cautionary Tales," 458.

are of much consequence, often drawing on crucial facts from ecological or socio-political worlds, and may be channelized to inculcate "a sense of responsibility for the future by attempting to prevent global catastrophes."[3]

Both of Smith's works under analysis in the present study can be categorized as 'narratives of entropy'—one fictional and the other non-fictional—that evince the gradual contamination of harmonious central ecosystems caused by anthropocentric interventions and the concomitant downfall of eco-centric habitats. Smith's short story *The Canker* signifies the power dynamics at work within the binaries involving exploitative systems and practices such as capitalism, colonization, androcentrism, and their devastating impacts on natural, indigenous, and feminocentric ecosystems. In her essay *Elegy for a Country's Seasons*, the author laments the disruption of the quotidian, as well as the psycho-emotional unnerving of the members of the populace, as she witnesses climate change in her own habitat in England. Zadie Smith disseminates caution in a voice marked by resolve and urgency, even as strategic solutions are explored.

Failing Ecosystems: The Indicative and the Real

Troubled by the state of a planet struggling on the cusp of chaos, major thinkers and activists of the twenty-first century have voiced concern and ire in the face of a rapidly worsening global ecological crisis. An all-pervading vocabulary to negotiate environmental calamity has formed over the decades, involving terms such as 'global warming,' 'climate change,' 'carbon footprint,' and the more recent 'pandemic,' and 'contagion'—grave yet indispensable buzzwords of the current moment. Not limited to a functional vocabulary that helps galvanize the worldwide movement for environmental preservation, the strength of narrative strategy, discourse, and rhetoric has been employed by leaders and thinkers the world over, to bring about an awakening to acute ecological crises, through authentic persuasion.

Almost a decade into the new millennium, in December 2009, the song "A Hard Rain's A-Gonna Fall," written by Nobel laureate Bob Dylan, was adopted by the United Nations as its unofficial anthem for the Copenhagen Climate Summit. The song rings true universally, with continuing relevance to situations and unwanted patterns in global climate change. The portentous lyrics of Dylan's song, which was first played in 1962, foretell scenarios of a deranged environment:

---

[3] Kurasawa, "Cautionary Tales," 455.

> I've stepped in the middle of seven sad forests
> I've been out in front of a dozen dead oceans
> [. . .]
> And it's a hard rain's a-gonna fall
> [. . .]
> I heard the sound of a thunder, it roared out a warnin'
> Heard the roar of a wave that could drown the whole world
> [. . .]
> Where the people are many and their hands are all empty
> Where the pellets of poison are flooding their waters
> [. . .]
> It's a hard rain's a-gonna fall.[4]

Barbara Plett's claim made earlier in the twenty-first century that Dylan's song "is now being invoked to highlight this generation's fear of environmental calamity,"[5] holds even more strongly today, as environmental damage continues to escalate.

However, nearly six decades after Dylan's poignant song debuted, contemporary British writer of the present moment, Zadie Smith in her *Elegy* complains of a lack of "intimate words"[6] to express the factual as well as emotive aspects of the escalating entropy caused by changing seasons, even though she observes the availability of "scientific and ideological language"[7] to delineate the same. Smith's essay addresses the entire scale of emotional and psychological responses of residents in England, towards disturbances in the expected seasonal patterns caused by global climate changes occurring in their local habitat. The disconcerted states of the residents of England as foregrounded by Smith are closely connected to terms from the field of mental health such as 'ecoanxiety' and 'solastalgia,' which signify the disturbed psycho-emotional states of individuals who fear the consequences of climate change. Castelloe describes ecoanxiety as an increasingly common and "a fairly recent psychological disorder,"[8] whereas Glenn Albrecht's term solastalgia

---

[4] Bob Dylan, "A Hard Rain's A-Gonna Fall," Bob Dylan, accessed March 9, 2021, http://www.bobdylan.com/songs/hard-rains-gonna-fall/.
[5] Barbara Plett, "Bob Dylan Song Adopted by Copenhagen Climate Summit," BBC News, updated December 5, 2009, http://news.bbc.co.uk/2/hi/8396803.stm.
[6] Zadie Smith, "Elegy for a Country's Seasons," in *Feel Free: Essays* (UK: Penguin Random House, 2018), 14.
[7] Ibid.
[8] Molly S. Castelloe, "Coming to Terms with Ecoanxiety," Psychology Today, January 9, 2018, https://www.psychologytoday.com/gb/blog/the-me-in-we/2018 01/coming-terms-ecoanxiety.

represents the specific psychological strain affecting persons whose geographical habitat undergoes environmental deterioration.[9] Zadie Smith's delineation of the conditions of varied ecosystems within her two narratives of entropy spans not just ecology, the individual self, and the associated socio-politics, but also their consequent disruption due to a breach of the laws and cycles of nature.

In Smith's allegorical parable *The Canker*, she brings forth the ecocritical aspects of 'contagion' affecting a fictional island inhabited by the central female protagonist Esorik, leading the social and biological ecosystems located in the island close to catastrophe. A work that is open to more than one critical interpretation, *The Canker* can be visualized as taking place in an imaginary past or alternately in an imaginary future; in a pre-developmental setting, or alternatively in a post-apocalyptic time. The narrative delineates the corruption of a matriarchal haven, by an androcentric and capitalist colonizer—an antagonist named The Usurper—who is the source of contagion for Esorik's island. The destruction of the island occurs because of multipronged causes that include the physical, the moral, the structural, and the political. Smith conveys the ruination unleashed within her story's setting through the use of powerful symbols from nature, depicting the marauding of biological resources.

The Canker and Smith's *Elegy* are discussed in this study, as 'dystopian imaginaries' and 'apocalypse narratives,' while emphasizing the dimensions of race, gender, and psychology, amidst the play of binary hierarchies formed by pairs such as androcentric/woman-centric, patriarchal/matriarchal, colonizer/native, savage/civilized, harmonious/ imbalanced and human/nature. Smith's two 'narratives of entropy'—the short fictional work *The Canker* and the essay *Elegy for a Country's Seasons*—that form the focus of this study may be examined as two narratives that expound crises in disparate types of ecosystems set across vastly different time-periods. Considered together, Smith's essay and her short story, imagine varied dimensions, scenarios, and futures for socio-biological structures that may be classified as 'failing ecosystems.' While the author makes a potent comment regarding the planet's depleting biological ecosystem in her *Elegy* essay, 'social ecosystems,' containing political, familial, and gendered sub-structures, are characterized in *The Canker*.

The deterioration towards catastrophe in each of the focal ecosystems is initiated by contact with certain pollutants and their subsequent spread

---

[9] Georgina Kenyon, "Have You Ever Felt 'Solastalgia'?," BBC Future, November 2, 2015, https://www.bbc.com/future/article/20151030-have-you-ever-felt-solastalgia.

through contagion, pollution, contamination, and infestation. At its simplest, pollution may be defined as "the action of polluting especially by environmental contamination with man-made waste."[10] It may be further emphasized that the word 'pollution,' "generally implies that the contaminants have an anthropogenic source—that is, a source created by human activities."[11] A closely related term, 'contagion' has been defined as "a corrupting or harmful influence that tends to spread; pollutant,"[12] "a corrupting influence or contact,"[13] or "an influence that spreads rapidly."[14] Other descriptions of 'contagion' denote it as "a disease-producing agent (such as a virus)."[15]

Each aspect of the terms 'pollution' and 'contagion,' may be interpreted as being at play in the narratives under scrutiny within the present study—ecologically, psychologically, or socio-politically—forming the basis of 'entropy' therein. Current real-world scenarios, when taken into account remind one that global crises such as the ongoing coronavirus pandemic are biologically rooted in the category 'contagion,' while also being closely related to environmental pollution and the ecological disasters that follow from it. Jonathan Watts underscores the view of Johan Rockström, director of the Potsdam Institute, that draws "a strong correlation between the pandemic and the environmental crisis,"[16] rooting the causation of pandemics in deforestation, wildlife trade, and air pollution, subsequently leading to weaker respiratory systems and the easier spread of viruses.

Smith's *The Canker* depicts mental and spiritual derangement as a metaphorical epidemic in an indigenous community, while her essay *Elegy for a Country's Seasons* fleshes out the 'lived experience' of England's inhabitants, drawing attention to the psychological, emotional, and ecological aspects of the ongoing disordering of seasons in the region.

---

[10] *Merriam-Webster.com Dictionary*, s.v. "pollution (n.)," accessed March 9, 2021, https://www.merriam-webster.com/dictionary/pollution.

[11] *Encyclopedia Britannica Online*, s.v. "Pollution," last modified February 11, 2021, https://www.britannica.com/science/pollution-environment.

[12] *Collins English Dictionary Online*, s.v. "contagion (n.)," accessed April 1, 2021, https://www.collinsdictionary.com/dictionary/english/contagion.

[13] *Merriam-Webster.com Dictionary*, s.v. "contagion (n.)," accessed April 1, 2021, https://www.merriam-webster.com/dictionary/contagion.

[14] Ibid.

[15] Ibid.

[16] Jonathan Watts, "Earth Day: Greta Thunberg Calls for 'New Path' after Pandemic," *The Guardian*, April 22, 2020, https://www.theguardian.com/environment/2020/apr/22/earth-day-greta-thunberg-calls-for-new-path-after-pandemic.

'Contagion,' in the context of the *Elegy* essay, refers to forms of environmental pollution impacting air, water, soil, flora, and fauna across the world, destroying the planet and its climate patterns. Smith indicates an impending catastrophe, denoting extreme weather events, and threatening ultimately to wipe out the entire humanity, as future generations approach their fruition. While the scope of the calamity depicted in Smith's *Elegy* is 'global,' the author chooses a 'local' apocalypse as the subject for her short story *The Canker*.

The Canker: Concentric Circles of Exploitation

Emphasizing the value of harmonious ecosystems in the tragic parable *The Canker*, Zadie Smith depicts the manner in which the disharmony in a human community translates into the loss of a pristine natural environment, as well as the violation of women and the endangered species. The narrative lends itself to specific threads of analysis in the human-nature hierarchy in ecocritical theory, concerning particularly the threat of androcentrism. Gender, civilization, and culture may be considered within the story, as conceptual refinements of the species category, and as aspects of the biodiversity concept. Smith's short story depicts the lives of a sheltered, indigenous island community that is woman-centric and lives in accordance with rhythmic cycles including "Labour, Praxis and Anima."[17] Bhutto perceives *The Canker* as depicting a 'cycle' as "a unit in the calendar used by a matriarchal society of Amazonians living on an island."[18] In addition to the cyclical pattern, asset-based division of labor is the norm among the women of the island, while men manage the domestic front. Esorik, a matriarch in her island community, an educated working-woman, and a mother, is the main protagonist of the story who makes a living by salting fish. She also teaches children how to tell stories and excels at it. Life on the island continues in perfect harmony until an evil masculine character known as the Usurper is "put in power"[19] by the misled inhabitants of the island.

The arrival of the Usurper causes a downward spiral in the minds of the people of the island, foreboding acute degeneration. Smith employs

---

[17] Zadie Smith,"The Canker," in *Grand Union: Stories* (UK: Penguin Random House, 2019), 209.

[18] Ali Bhutto, "Book Review: Zadie Smith's 'Grand union' Gets Better with Every Read," *The Wire*, November 14, 2019, https://thewire.in/books/book-review-zadie-smiths-grand-union-gets-better-with-every-read.

[19] Smith,"The Canker," 208.

a powerful symbol from nature, that of "The Canker in the Rose,"[20] to elucidate the manner in which the Usurper's dominion infects and destroys the original pristine balance characterizing the island's people. The ugliness and perversity of the Usurper's mind, soon runs rife through both the inner and outer selves of Esorik's people like a silent blight, just as a canker destroys and decays a rose from the very inside. The Usurper, a man from the mainland may be seen as representative of aspects of the world that are autocratic, perverse, and possibly imperialist. Esorik and her kind, on the other hand, seem to constitute sections of a population that are naturally harmonious, spiritual, indigenous, and woman-centric. Zadie Smith states her motivation and the real-world basis for introducing the Usurper as a character in *The Canker*, in an interview on the subject of her short story collection *Grand Union: Stories (2019)*,

> We have plenty of usurpers right now. [. . .] They're all over the planet. But I was just interested in the idea of the bully—particularly the masculine bully, who cast his shadow over a people. And I wanted to try and imagine that world in its bare bones.[21]

Viewed as an ecocritical text, *The Canker* may be deconstructed to enable the classification of the imperialist and imperialized characters depicted within the narrative as structural divisions within the 'race' concept. Furthermore, 'race' itself may be categorized as a factor pertaining to the biological 'species' concept, while feminocentrism and androcentrism signify aspects of the 'gender' concept. The categories of gender and race also hold within their domains, divisions that reveal aspects of biodiversity. Smith's short story then, is an allegory for the invasion of a civilized feminocentric indigenous society by a barbaric androcentric dictatorial ruler.

As an accomplished woman, as well as a wise and loyal member of her community, Esorik senses danger even initially, in opening doors to the Usurper. Once in possession of clear evidence of the Usurper's misogyny and his disregard of the indigenous cycles, rhythms, and traditions of Esorik's community, the protagonist is offended and furious. She then enters the former's mind using a mystical gift she possesses,

---

[20] Smith,"The Canker," 211.
[21] Lynn Neary, "Zadie Smith Has an Eclectic Mix of Short Stories in 'Grand Union'," NPR, October 5, 2019, https://www.npr.org/2019/10/05/767488497/zadie-smith-has-an-eclectic-mix-of-short-stories-in-grand-union.

making the ugliness inside the Usurper's mind common knowledge on the island. The island's people now start mimicking and enjoying the Usurper's barbarism, albeit in mockery. Yacovissi, in an analysis of Smith's present short story, draws attention to an essential quality of the Usurper which leads to his gaining total power over Esorik's community—"his greatest disruptive power is to make himself the universal focus of all attention, which makes everyone, [. . .] complicit in the debasement of social norms."[22]

In order to convey the spiritual and intellectual deterioration of the people of Esorik's island, the author presents the image of the hunting of the white hart. This image depicts the cold-blooded plundering of one of nature's most rare gifts by the once harmonious islanders. In their mad hatred for the Usurper, Esorik's people lose their sanity as well as humanity, slaying innocent white harts for sport,

> Hunting the white hart returned to fashion. Friends of Esorik, [. . .] chased that elusive animal through its habitat, stabbed it in multiples of seven—seven strokes each—and then stood over the poor beast's heaving belly as it bled out into the earth and over their shoes.[23]

Members of the island community are driven to slaughtering the rare and guileless animal due to their loathing for the Usurper, who believes that the white hart is the source of his power. It is the island's natural ecosystem then that bears the brunt of the Usurper's sway deep within the selves of the islanders.

The Usurper's invasion of Esorik's island draws a parallel with Europe's vast imperialism over races including various indigenous peoples and native tribes. Scenarios of American capitalism, the accompanying disruption of the Earth's natural harmony, and the natural habitats of endemic species are also reflected in the perversion of the island's people and the subsequent killing of the white hart. Similarly, Smith's use of the symbol of the canker in the rose to represent the pollution and disease brought about by a contagion of foreign origin may be interpreted as signifying exploitative real-world systems and phenomena.

---

[22] Jennifer Bort Yacovissi, "'Grand Union Stories' – A Book by Zadie Smith," *The Lighthouse Peddler*, January 1, 2020,
https://www.thelighthousepeddler.com/archive-2018/2020/1/grand-union-stories-a-book-by-zadie-smith-reviewed-by-jennifer-bort-yacovissi.
[23] Smith, "The Canker," 211.

The Usurper comes to Esorik's island as a polluter, who desires the endless expansion of his powers, without any regard for the welfare of his subjects. Critics have equated the character of the Usurper with several major leaders of nations across time including "Assad, Trump, Erdogan, Urban or any other."[24] The Usurper may thus be viewed as a colonizer or a capitalist trader seeking to multiply his own power by exploiting the resources intrinsic to a previously unexplored territory, resulting in its environmental degradation. As an act of violation of the sanctity of an ecosystem, and a second instance of the endangerment of an endemic species in *The Canker*, the Usurper's destruction of the clan of the Ekalbia parallels systematic racial selection under colonization or capitalism. The Ekalbia may be seen as an ethnic group, a tribe, or a variation of the race concept and their elimination may be interpreted as being analogous to the wiping out of an entire species.

The extermination of the Ekalbia closely resembles environmental poisoning/pollution or the poaching of the natural world for money or power. Esorik anticipates an imminent apocalypse making its way towards her island community when she sees them imbibing the nature of the Usurper, just as a canker works its way to destroy a pristine rose that symbolizes thriving nature. Malec proposes that Smith's short story is a fable set in "a post-apocalyptic community,"[25] led by a man named the Usurper. Malec's observation suggests that apocalypse befalls Esorik's island just as the Usurper arrives, and is in full progress by the time the narrative reaches its culmination. Esorik wishes for her fellow indigenous natives to remain free of the influence of the Usurper, who represents imperial, capitalist, tyrannical, pollutant, or androcentric powers. She is wary of all those who disrupt natural cycles and rhythms, spreading depravity and hatred in their place.

From an ecofeminist perspective, the symbolization of nature being ravaged in *The Canker*, including 'the hunting of the white hart' and 'the canker in the rose,' runs parallel to a decline in the stature of women in Esorik's community. The Usurper's 'circle' consists only of women whom he has complete power over. As he looms over the body of a young girl in

---

[24] The Newsroom, "Book Review: Grand Union: Stories, by Zadie Smith," *The Scotsman*, October 20, 2019, https://www.scotsman.com/news/book-review-grand-union-stories-zadie-smith-2501118.

[25] Jennifer Malec, "Fragments, Explorations and Variations," *The Johannesburg Review of Books*, November 4, 2019, https://johannesburgreviewofbooks.com/2019/11/04/fragments-explorations-and-variations-jennifer-malec-reviews-zadie-smiths-debut-collection-of-short-stories-grand-union-her-most-american-book-to-date/.

a public ritual, he makes an address described as "The Father Who Eats His Young,"[26] suggesting the possibility of incest. Conversely, the women on Esorik's island are described as having built bridges, designed civic systems, sat on justice committees, and danced joyously through the streets before coming under the influence of the Usurper.[27] These occupations represent the struggles that the island's women engage in for survival and for nurturing their people, as well as their celebration of selfhood in a feminocentric society. The prominent ecofeminist critic Vandana Shiva observes in her work *Staying Alive: Women, Ecology and Development*, "In the perspective of women engaged in survival struggles which are, simultaneously, struggles for the protection of nature, women, and nature are intimately related, and their domination and liberation similarly linked."[28] It is ironical then, that the very women of Esorik's island who once achieved laurels and strengthened their community are reduced to "cackling and whooping"[29] in a degenerate manner, as they repeat the ugly "songs and riddles, filthy jokes and rhyming curses"[30] originating from the Usurper. The demolition of nature and degradation of the position of women within the narrative are reflected in each other.

Disrupted Seasons and Unsettled Selves: Smith's Contemporary England

The narrative deconstruction of *The Canker* unveils the ecological and socio-political implications of devolving ecosystems impacting the spheres of gender, species extinction, and racial selection. Contemplating the human causes and responses associated with climatic deviations, Zadie Smith in the essay *Elegy for a Country's Season's* delineates the psycho-social effects of an environmental crisis caused by the irresponsible choices of modern society. In a personal rendition of the distortion of seasonal experiences in England, her lifelong ecological home and habitat—Smith's *Elegy* takes head-on, the urgent issue of global climate change. The author considers in specific detail the scientific, social, cultural, and emotional outcomes of ecological imbalances in the territory

---

[26] Smith,"The Canker," 210.
[27] Smith,"The Canker," 208.
[28] Vandana Shiva, *Staying Alive: Women, Ecology and Survival in India*, (New Delhi: Kali for Women, 1988), 45,
https://ia600408.us.archive.org/4/items/StayingAlive-English-VandanaShiva/Vandana-shiva-stayingAlive.pdf.
[29] Smith,"The Canker," 211.
[30] Smith,"The Canker," 210.

of England, making connections with worldwide irregularities that are presently on the rise. Smith identifies patterns among a wide spectrum of human psycho-emotional responses to the growing visible signs of altered climatic manifestations.

A critical commentator of the ecology and the pandemic, the author has remained close to England over the decades—her place of birth and upbringing, as well as the place she calls home. Hadjetian states that, "Zadie Smith was born in the north-west London borough of Brent in 1975. [. . .] she grew up in the multicultural community of Willesden Green and still lives in her home city."[31] Smith herself refers to London as her home when she states, "I work in an American university, but in the long academic holidays I come home, to England."[32] The author emphasizes in her interview to Chotiner that she "always thought of England as home."[33] The increasing uneasiness affecting the residents of Smith's home territory results from the dichotomy between the quintessential state of its seasonal cycle and the 'abnormal' state that replaces it. Disturbing emotions often develop due to this undesirable shift as well as the loss of relatable surroundings, and may include sadness, guilt, and shame. The author connects these emotions to the citizens' recurring use of the euphemism "the new normal,"[34] which is employed by them to describe the new and unwelcome environmental scenario, while overlooking the necessity of acknowledging the distasteful truth of their present reality. This betrays a recently manifested reluctance on the part of the residents regarding harking back to a natural normalcy that remains "painful to remember."[35]

Smith also draws attention to shock and anger as other emotional signs of climate change. Some citizens, as Smith notes, adopt the emotional coldness typical of pragmatism by focusing only on immediate survival and personal profiteering—a sign of negligence fatal to both the

---

[31] Sylvia Hadjetian, *Multiculturalism and Magic Realism in Zadie Smith's Novel White Teeth: Between Fiction and Reality* (Hamburg: Anchor Academic Publishing, 2015), 8.
[32] Zadie Smith, "Zadie Smith on Her Markedly Different Style in London versus New York," *Vogue*, October 27, 2019, https://www.vogue.co.uk/fashion/article/zadie-smith-london-new-york-fashion.
[33] Isaac Chotiner, "Critics, Appropriation, and What Interests Her Novelistically about Trump," *Slate*, November 16, 2016, https://slate.com/culture/2016/11/a-conversation-with-zadie-smith-about-cultural-appropriation-male-critics-and-how-trump-interests-her-novelistically.html.
[34] Smith, "Elegy for a Country's Seasons," 14.
[35] Ibid.

immediate and wider ecosystems. Other responses of individuals include choosing outright diversion or adopting a coping mechanism. The teenaged Swedish climate activist Greta Thunberg takes an unusual approach in calling for allowing an unhindered escalation of inner disturbances until the climate crisis can no longer be ignored, so that it leads to relevant action. In a speech given at Davos, Thunberg employs a metaphor of violence and loss to convey to all 'adults,' "I don't want you to be hopeful. I want you to panic. [. . .] And then I want you to act. I want you to act as you would in a crisis. I want you to act as if our house is on fire. Because it is."[36]

Clayton et al., in their report titled *Mental Health and Our Changing Climate* produced for the American Psychological Association (APA), provide a conceptual framework that validates the psycho-emotional repercussions of England's changing seasonal patterns, as delineated by Zadie Smith. The APA report introduces the term 'ecoanxiety,' which consolidates a wide gamut of stressful emotional responses displayed by individuals in reaction to current or anticipated environmental alterations worldwide. Clayton et al. make the following deductions regarding the psycho-emotional manifestations of climate change:

> Gradual, long-term changes in climate can also surface a number of different emotions, including fear, anger, feelings of powerlessness, or exhaustion [. . .] Watching the slow and seemingly irrevocable impacts of climate change unfold, and worrying about the future for oneself, children, and later generations, may be an additional source of stress [. . .] Albrecht (2011) and others have termed this anxiety ecoanxiety.[37]

Zadie Smith analyzes the climatic mayhem adrift in her part of the world, not only as a seasoned intellectual but also as a sensitive human who displays signs of ecoanxiety herself. The author laments the loss of the predetermined normal, and certain "seemingly small things,"[38] comprising features of the environmental heritage of England. Therefore,

---

[36] Greta Thunberg, "'Our House Is on Fire': Greta Thunberg, 16, Urges Leaders to Act on Climate," *The Guardian*, January 25, 2019, https://www.theguardian.com/environment/2019/jan/25/our-house-is-on-fire-greta-thunberg16-urges-leaders-to-act-on-climate.
[37] Susan Clayton et al., *Mental Health and Our Changing Climate: Impacts, Implications and Guidance*, (Washington, D.C.: American Psychological Association, and ecoAmerica, 2017), 27, https://www.apa.org/news/press/releases/2017/03/mental-health-climate.pdf.
[38] Smith, "Elegy for a Country's Seasons," 15.

ecoanxiety characterizes the emotional states of the members of the populace, and also of the author herself. The concept of 'solastalgia' defined by Clayton et al. as given below is a related term that problematizes Smith's individual and social angst concerning her 'home environment' or 'natural habitat':

> As climate change irrevocably changes people's lived landscapes, large numbers are likely to experience a feeling that they are losing a place that is important to them—a phenomenon called solastalgia. [. . .] a sense of desolation and loss similar to that experienced by people forced to migrate from their home environment.[39]

Smith's *Elegy* may be seen to stem from a keen feeling of 'solastalgia,' and the accompanying dismay at its core.

Dissonant Cycles: Earth, Woman, Humanity

As discussed previously, the growing mental and spiritual dissonance within the Usurper-hating people of Esorik's island mirrors the detestable nature of the Usurper himself. A similar dissonance is recognized by Smith in her *Elegy* essay, in the human reactions to changing seasons caused in turn by a deteriorating natural environment. It has been seen that disharmony between human beings and their natural ecology, as well as imbalances created by humans within nature, lead to a lack of balance within the individual over time. A lack of inner consonance is perceptible in the common plight of Esorik's islanders and the ecoanxiety of the people of Smith's England. The Australian environmental thinker Elyne Mitchell had warned readers even in 1946, of the pitfalls of allowing the equation between Earth's cycles and humanity to be damaged. Mitchell states the significance of the organic link between mankind and the ecosystems it inhabits, accentuating that, "the break in this unity is swiftly apparent in the lack of 'wholeness' in the individual person."[40] Schlanger

---

[39] Clayton et al., *Mental Health and Our Changing Climate*, 25.
[40] Elyne Mitchell, *Soil and Civilization*, (Sydney: Halsted Press, 1946), 4, **quoted in** Zoë Schlanger, "A Philosopher Invented a Word for the Psychic Pain of Climate Change," *Quartz*, October 13, 2018, https://qz.com/1423202/a-philosopher-invented-a-word-for-the-psychic-pain-of-climate-change/amp/.

foregrounds Mitchell's emphasis on the detrimental social impact caused by the loss of humanity's "stable bond to Earth's cycles and systems."[41]

Drawing attention to the findings of the American Psychological Association (APA) pertaining to the effects of climate change on indigenous peoples, Schlanger states that, "More often than not [. . .] people of colour, and indigenous people will feel these impacts first, and in many cases, already are, the APA writes. The impacts of climate change are not distributed equally."[42] The roots of this disparity are possibly related to the indiscriminate physical exploitation of the homelands of people of colour and indigenous people through colonization and capitalism by the Euro-American bloc of the world nations. Zadie Smith indicates the greater impact of climate change on people of colour in her essay, elaborating on the major climatic occurrence of Superstorm Sandy—a recent "massive storm that brought significant wind and flooding damage [. . .] in late October 2012" to areas including the Caribbean and the United States.[43] Smith emphasizes that the storm first hit Jamaica, which is home to a great many people of colour. She also foregrounds the increased likelihood of certain places inhabited by coloured races and ethnicities attracting an environmental apocalypse, stating in her essay,

> in Jamaica, where Sandy first made landfall, the ever more frequent tropical depressions, storms, hurricanes, droughts and landslides do not fall, for Jamaicans, in the category of ontological argument. [. . .] The apocalypse is always usefully cast into the future—unless you happen to live in Mauritius, or Jamaica, or the many other perilous spots.[44]

It must be noted that the Usurper is a character who ridicules "the very concept of the cycle."[45] The concept of the cycle may be associated with the female body, with the seasonal cycles, and with the three rhythmic cycles of 'Labour, Praxis and Anima,' by which the women of

---

[41] Zoë Schlanger, "A Philosopher Invented a Word for the Psychic Pain of Climate Change," *Quartz*, October 13, 2018, https://qz.com/1423202/a-philosopher-invented-a-word-for-the-psychic-pain-of-climate-change/amp/.

[42] Zoë Schlanger, "We Need to Talk about 'Ecoanxiety': Climate Change Is Causing PTSD, Anxiety, and Depression on a Mass Scale," *Quartz*, April 4, 2017, https://qz.com/948909/ecoanxiety-the-american-psychological-association-says-climate-change-is-causing-ptsd-anxiety-and-depression-on-a-mass-scale/.

[43] *Encyclopedia Britannica Online*, s.v. "Superstorm Sandy," last modified March 2, 2021, https://www.britannica.com/event/Superstorm-Sandy.

[44] Smith, "Elegy for a Country's Seasons," 17.

[45] Smith, "The Canker," 210.

the island have lived. The androcentric Usurper, by insulting the concept of the cycles displays, not only hatred for women, but also a disregard for nature. Placing this framework within the scenario of the planet Earth, it may be seen that gross violations of natural rhythms and cycles over recent centuries have led to extreme weather events closer to the present. These include Superstorm Sandy, which was characterized as "a raging freak of nature"[46] by the media; owing to its sheer impact and the destruction it caused.

Shiva contends that "it is human action, it is human greed, it is human sin and stupidity that is making Mother Nature react,"[47] suggesting a rationale for extreme weather events and diseases such as epidemics. The current state of the global biological environment shows clear signs of abuse inflicted through aberrant tendencies such as those put forth by Shiva. The disruption of natural rhythms affected by the rule of the Usurper leads to the development of unnatural and destructive tendencies within Esorik's people, eventually causing a calamity that is to the detriment of women, nature, and children of the island, as well as the Ekalbia. On considering Smith's *Elegy* in parallel, it may be seen that an impending catastrophe awaits the modern world, as explicated by the author's description of irregular seasonal cycles and extreme weather events.

Conclusion: From Elegy to Caveat

Zadie Smith's twenty-first century narratives *The Canker* and *Elegy for a Country's Seasons* may be seen as cautionary discourses that each envisage a fast approaching apocalypse. The creative writer and environmental activist Margaret Atwood suggests replacing the term climate change with the phrase "everything change," owing to the extensive impact of climate change on oxygen and water-dependent species, as well as its influence on rainfall, agriculture, housing, livestock, and wildlife.[48] Atwood's exposition relates the dependence of different species on the vagaries of

---

[46] Sarah Gibbens, "Hurricane Sandy, Explained," *National Geographic*, February 12, 2019, https://www.nationalgeographic.com/environment/article/hurricane-sandy.

[47] Vandana Shiva, "Dr Vandana Shiva, Rights of Nature and Earth Democracy," Global Alliance for the Rights of Nature, March 25, 2014, YouTube video, 00:13:32, https://www.youtube.com/watch?v=AhrmgUdfO44.

[48] Margaret Atwood, "An Interview with Margaret Atwood," interview by Ed Finn, *Slate*, February 6, 2015, https://slate.com/technology/2015/02/margaret-atwood-interview-the-author-speaks-on-hope-science-and-the-future.html.

the climate, to oxygen and water, which are sourced only through environmental forces.

In his essay titled "The Importance of Apocalypse," Schatz discusses the types of discourses in various literary forms on the theme of environmental apocalypse, and their seminal role in mobilizing initiative, as part of the practice of ecocriticism.[49] While Zadie Smith, examining the journey of humanity thus far in her *Elegy* essay, deduces that, "the terrible truth is that we had a profound, historical attraction to apocalypse,"[50] Schatz contests this position, arguing, "Oftentimes it takes images of planetary annihilation to motivate people into action after years of sitting idly by watching things slowly decay [. . .] to compel policymakers to enact even piecemeal reform."[51]

Smith's essay questions the convenient division of all persons affected by climate change, into two categories titled "alarmists" and "realists."[52] However, it remains hard to categorize her *Elegy* as being one that seeks to raise an alarm. Even if Smith's essay suggests an ecological crisis at hand, her focus is to raise awareness and motivate responsible initiative by citizens. This non-fictional apocalyptic narrative is constructive, not alarmist. However, *The Canker* is a more poignant parable that is disquieting in effect, given its racial, gendered, ecological, and planetary implications in modern times. *The Canker* and *Elegy for a Country's Seasons* may be seen as dystopian imaginaries that bring out the irony in the present climate change scenario, encouraging a preemptive approach to environmental conservation through ecological responsibility on a global scale, for the prevention of natural disasters and other climatic aberrations. Kurusawa underscores the functionality of the notion of a 'dystopian imaginary,' opining that,

> instead of bemoaning the contemporary preeminence of a dystopian imaginary [. . .] it can enable a novel form of transnational socio-political action [. . .] that can be termed preventive foresight. [. . .] [I]t is a mode of ethico-political practice enacted by participants in the emerging realm of global civil

---

[49] Joseph Leeson Schatz, "The Importance of Apocalypse: The Value of End-of-the-World Politics while Advancing Ecocriticism," *The Journal of Ecocriticism* 4, no. 2 (July 2012): 20-33, https://ojs.unbc.ca/index.php/joe/article/view/394/382.
[50] Smith, "Elegy for a Country's Seasons," 19.
[51] Schatz, "The Importance of Apocalypse," 21.
[52] Smith, "Elegy for a Country's Seasons," 17.

society [. . .] [by] putting into practice a sense of responsibility for the future by attempting to prevent global catastrophes.[53]

In its conclusion, Smith's essay steers away from panic or lamentation in its tone, pointing instead towards a practical approach as the solution, moving from the "elegiac" to the "practical."[54] The author delivers the message of the purposelessness of psycho-emotional chaos, suggesting instead the merits of choosing science, integrity, and strategy as means to attain ecological security – a shift she suggests from "what have we done?" to "what can we do?"[55] The fictional work *The Canker*, on the other hand, is focalized around the consciousness of Esorik, the central female protagonist. Even though a man (the Usurper) is the political ruler in Smith's story, it is the woman's (Esorik's) perspective and empathy that is conveyed throughout. An allegorical parable replete with symbolism, the story is also a stark tale of the deterioration of a race or tribe, imparting epic dimensions. A narrative of the devastation of biological and social ecosystems to the point of a near apocalypse, it envisions possibilities, albeit delayed, for rebuilding an invaded civilization after it has experienced moral and structural ruination. Finally, *The Canker* is both a puzzle open to exegesis, as well as a caveat for humanity intended to prevent the calamities that cause the failure of ecosystems and those that stem from taking life, nature, freedom, and dignity for granted.

References

Atwood, Margaret. "An Interview with Margaret Atwood." By Ed Finn. *Slate*, February 6, 2015. https://slate.com/technology/2015/02/margaret-atwood-interview-the-author-speaks-on-hope-science-and-the-future.html.

Bhutto, Ali. "Book Review: Zadie Smith's 'Grand union' Gets Better with Every Read." *The Wire*. November 14, 2019. https://thewire.in/books/book-review-zadie-smiths-grand-union-gets-better-with-every-read.

---

[53] Fuyuki Kurasawa, "Cautionary Tales: The Global Culture of Prevention and the Work of Foresight," *Constellations* 11, no. 4 (2004): 454-455, **quoted in** Joseph Leeson Schatz, "The Importance of Apocalypse: The Value of End-of-the-World Politics while Advancing Ecocriticism," *The Journal of Ecocriticism* 4, no. 2 (July 2012): 21-22, https://ojs.unbc.ca/index.php/joe/article/view/394/382.
[54] Smith, "Elegy for a Country's Seasons," 19.
[55] Ibid.

Castelloe, Molly S. "Coming to Terms with Ecoanxiety." *Psychology Today*. January 9, 2018. https://www.psychologytoday.com/gb/blog/the-me-in-we/201801/coming-terms-ecoanxiety.

Chotiner, Isaac. "Critics, Appropriation, and What Interests Her Novelistically about Trump." *Slate*, November 16, 2016. https://slate.com/culture/2016/11/a-conversation-with-zadie-smith-about-cultural-appropriation-male-critics-and-how-trump-interests-her-novelistically.html.

Clayton, Susan, Christie Manning, Kirra Krygsman, and Meighen Speiser. *Mental Health and Our Changing Climate: Impacts, Implications and Guidance*. Washington, D.C.: American Psychological Association, and ecoAmerica, 2017. Accessed February 12, 2021. https://www.apa.org/news/press/releases/2017/03/mental-health-climate.pdf.

Collins English Dictionary Online, s.v. "contagion (n.)," Accessed April 1, 2021. https://www.collinsdictionary.com/dictionary/english/contagion.

Dylan, Bob. "A Hard Rain's A-Gonna Fall." Bob Dylan. Accessed March 9, 2021. http://www.bobdylan.com/songs/hard-rains-gonna-fall/.

Encyclopedia Britannica Online, s.v. "Pollution," last modified February 11, 2021. https://www.britannica.com/science/pollution-environment.

Encyclopedia Britannica Online, s.v. "Superstorm Sandy," last modified March 2, 2021. https://www.britannica.com/event/Superstorm-Sandy.

Gibbens, Sarah. "Hurricane Sandy, Explained." *National Geographic*. February 12, 2019. https://www.nationalgeographic.com/environment/natural-disasters/reference/hurricane-sandy/.

Hadjetian, Sylvia. *Multiculturalism and Magic Realism in Zadie Smith's Novel White Teeth: Between Fiction and Reality*. Hamburg: Anchor Academic Publishing, 2015.

Kenyon, Georgina. "Have You Ever Felt 'Solastalgia'?" BBC Future. November 2, 2015. https://www.bbc.com/future/article/20151030-have-you-ever-felt-solastalgia.

Kurasawa, Fuyuki. "Cautionary Tales: The Global Culture of Prevention and the Work of Foresight." *Constellations* 11, no. 4 (November 2004): 453-475. https://doi.org/10.1111/j.1351-0487.2004.00389.x.

Malec, Jennifer. "Fragments, Explorations and Variations – Jennifer Malec Reviews Zadie Smith's Debut Collection of Short Stories, Grand Union, Her Most American Book to Date." *The Johannesburg Review of Books*, November 4, 2019.

https://johannesburgreviewofbooks.com/2019/11/04/fragments-explorations-and-variations-jennifer-malec-reviews-zadie-smiths-debut-collection-of-short-stories-grand-union-her-most-american-book-to-date/.

Merriam-Webster.com Dictionary, s.v. "contagion (n.)," Accessed April 1, 2021. https://www.merriam-webster.com/dictionary/contagion.

Merriam-Webster.com Dictionary, s.v. "pollution (n.)," Accessed March 9, 2021. https://www.merriam-webster.com/dictionary/pollution.

Neary, Lynn. "Zadie Smith Has an Eclectic Mix of Short Stories in 'Grand Union'." NPR. October 5, 2019. https://www.npr.org/2019/10/05/767488497/zadie-smith-has-an-eclectic-mix-of-short-stories-in-grand-union.

The Newsroom. "Book Review: Grand Union: Stories, by Zadie Smith." *The Scotsman*, October 20, 2019. https://www.scotsman.com/news/book-review-grand-union-stories-zadie-smith-2501118.

Plett, Barbara. "Bob Dylan Song Adopted by Copenhagen Climate Summit." BBC News. updated December 5, 2009. http://news.bbc.co.uk/2/hi/8396803.stm.

Schatz, Joseph Leeson. "The Importance of Apocalypse: The Value of End-of-the-World Politics while Advancing Ecocriticism." *The Journal of Ecocriticism* 4, no. 2 (July 2012): 20-33. https://ojs.unbc.ca/index.php/joe/article/view/394/382.

Schlanger, Zoë. "A Philosopher Invented a Word for the Psychic Pain of Climate Change." *Quartz*. October 13, 2018. https://qz.com/1423202/a-philosopher-invented-a-word-for-the-psychic-pain-of-climate-change/amp/.

Schlanger, Zoë. "We Need to Talk about 'Ecoanxiety': Climate Change Is Causing PTSD, Anxiety, and Depression on a Mass Scale." *Quartz*. April 4, 2017. https://qz.com/948909/ecoanxiety-the-american-psychological-association-says-climate-change-is-causing-ptsd-anxiety-and-depression-on-a-mass-scale/.

Shiva, Vandana. "Dr Vandana Shiva, Rights of Nature and Earth Democracy." Global Alliance for the Rights of Nature. March 25, 2014. YouTube video, 00:13:32. https://www.youtube.com/watch?v=AhrmgUdfO44.

Shiva, Vandana. *Staying Alive: Women, Ecology and Survival in India*. New Delhi: Kali for Women, 1988. Accessed February 10, 2021. https://ia600408.us.archive.org/4/items/StayingAlive-English-VandanaShiva/Vandana-shiva-stayingAlive.pdf.

Smith, Zadie. "The Canker." In *Grand Union: Stories*, 207–212. UK: Penguin Random House, 2019.

Smith, Zadie. "Elegy for a Country's Seasons." In *Feel Free: Essays*, 14-19. UK: Penguin Random House, 2018.

Smith, Zadie. "Zadie Smith on Her Markedly Different Style in London versus New York." *Vogue*, October 27, 2019. https://www.vogue.co.uk/fashion/article/zadie-smith-london-new-york-fashion.

Sontag, Susan. "The Imagination of Disaster." In *Against Interpretation and Other Essays*, 212-228. New York: Dell Publishing, 1961.

Thunberg, Greta. "'Our House Is on Fire': Greta Thunberg, 16, Urges Leaders to Act on Climate." *The Guardian*, January 25, 2019. https://www.theguardian.com/environment/2019/jan/25/our-house-is-on-fire-greta-thunberg16-urges-leaders-to-act-on-climate.

Watts, Jonathan. "Earth Day: Greta Thunberg Calls for 'New Path' after Pandemic." *The Guardian*, April 22, 2020. https://www.theguardian.com/environment/2020/apr/22/earth-day-greta-thunberg-calls-for-new-path-after-pandemic.

Yacovissi, Jennifer Bort. "'Grand Union Stories' – A Book by Zadie Smith." *The Lighthouse Peddler*, January 1, 2020. https://www.thelighthousepeddler.com/archive-2018/2020/1/grand-union-stories-a-book-by-zadie-smith-reviewed-by-jennifer-bort-yacovissi.

CHAPTER FIFTEEN

HAVE WE LOST OUR SENSES?

ROMA MADAN-SONI

In the month that I begin to write about climate change, my body begins to sweat. I wake in the night, flooded with heat. ... The heat pours down my neck and shoulders, arms, spine, leaving me sweating, then chilled. Nothing is wrong. This sudden heat is a step toward my own mortality, natural and inevitable. I think about what it is like to be a body, overheating.[1]

Introduction

The world that environs us, that is around us, is also within us.
    We are made of it; we eat, we drink, and breathe it; it is bone of our bone and flesh of our flesh.[2]

The Sundarbans was never an easy place to live since fresh water was available only during the monsoons.[3] The thick mangroves covering the sprawling delta swarmed with wildlife, with many species fatal for humans. The indigenous Adivasis lived in conjunction with land, water, and habitat species. Kolkata was established as the major riverine port of the Indian subcontinent by the British in the mid-18th century when they sailed down rivers thickly lined with primeval-indigenous mangroves. Subsequently, the Sundarbans were colonized. Small earthen embankments replaced gatekeeper mangroves whose deep-reaching roots and oxygen diffusing branches were hacked for timber; a tract system was set to civilize the 'wild' land as the colony's food bank. Cyclones, hurricanes, wildfires, climate change, fertilizers, and pesticides then began to invade and erode the land

---

[1] Gaard, 2017.
[2] Berry, 1993
[3] Stone and Hornak, 2019.

and its native species, spinning them into migrants in India's cities as slum dwellers.

We systematically ignore nature's nurturing and protective role, retaining it within a politics of invisibility and subservience.[4] A "forgotten space"[5] that is cast off. The 'anthropogenic-cene eye' rarely notices the context of the submerged ecological upheaval that manifests in the form of a steady rise in sea levels, together with the experiential "Slow Violence"[6] of climate crisis amongst aquatic species and eco-materialisms. This would go unseen, misunderstood, and abandoned if it did not trigger the death of the planet's largest mangrove forest that combats global warming by absorbing rather than emitting carbon dioxide, creating a carbon sink.[7] Furthermore, "It's no consequence we treat migrants like dirt,"[8] and their contribution within the community goes unnoticed. Earthen dikes buttressed with mangrove restitution could work as a better strategy than all those undertaken in the past. "We have passports, we have political borders, but there's no atmospheric border," warns Hazra, "It is not a national issue anymore. It is transnational."[9]

> "Art has a unique power to transcend differences and connect with people on a visceral level – and compel action."[10]

Over the past two decades, a profusion of live cultural artworks, performances, screenings, and fictional texts have shown the amplified humanoid impact on climate change in local and global arenas. Dan Bloom's catchy phrase 'cli-fi' bridges the gap between academia and popular culture to construct a blended language, emerging in publications and academic conferences, and materializing in promotional undertakings, such as the emergence of the first how-to guides like Szabo's *Saving the World One Word at a Time: Writing Cli-Fi*[11], the institution of Brady's monthly column, "Burning Worlds."[12]

---

[4] Buchanan and Jeffery, 2019.
[5] Sekula, 2010.
[6] Nixon, 2013.
[7] Akhand, 2021.
[8] Gaard, 2017.
[9] Hazra, in Stone and Hornak, 2019.
[10] Jay, 2019
[11] Szabo, 2015.
[12] Brady, 2020.

Hybrid Ecologies

> I am not talking about inventing fairies at the bottom of the garden. It's a matter of being open to experiences of nature as powerful, agentic, and creative, making space in our culture for an animating sensibility and vocabulary.[13]

In this chapter I show how, in the same way we employ the medium of our sensory systems to communicate in the world of the womb, after our birth we need to reconnect with the world's Ecosystem, our home. Our senses will help us build a just living world, one more closely adapted to life: to our interbeing with humans of varied genders, colours, races, and nations, and with different species and features of this one priceless planet. Usually, cli-fi, climate justice narratives, and art align and are birthed from the same womb but are published separately. Through the medium of a series of my paintings in 2020 that trigger our senses and emotions, and the supporting text based on scientific analysis, we can connect with planetary sensations of earth-others; together they strengthen my narrative. Through my artistic explorations and writing, I investigate human conception and perception of self-identity as rooted within the kin-centric[14] and branched in their loci as intra-active[15] towards trans-species, ecological and social justice.

Our physical identities are envisaged as material flows rather than distinct and confined bodies, and accordingly, we are required to "reimagine 'climate-change' and the fleshy, damp immediacy of our own embodied existences as intimately imbricated."[16] "The weather and climate are not phenomena 'in' which we live at all—where climate would be some natural backdrop to our separate human dramas—but are rather of us, in us, through us."[17] This trans-corporeality is "embodied, but never essentialized;" it is always voiced "in ways that cannot be dissociated from politics, economics, coloniality and privilege—and my [OUR] embeddedness therein."[18]

We have no time to experience cli-fi effects and take action. The criticality of climate change urges me to source from global mythologies that have long associated women with an ocean-animal form. Along coasts of Africa and the Americas, the woman is worshiped as the dark-blue-

---

[13] Plumwood, 2009.
[14] Salmon, 2011.
[15] Barad, 2007.
[16] Neimanis and Walker, 2014.
[17] Ibid. 2014, 559.
[18] Neimanis, 2013.

fishtailed Olokun who wears a deep-sea-coral necklace; she conglomerates 'Coral Wombs' as Mother of Fishes. Embodied as Syrian Mermaid-Goddess Atargatis, Aztec-deity Chalchiuhtlicue, Australian-aboriginal Enigma, Iniut- Sedna, Indian-Goddesses Ganga-Yamuna or Irish merman, her dual role is associated with womb and water. This human-animal embodiment reflects the weight and roots of diverse cultural traditions. As symbols of strength, fertility, and regeneration, they are an enduring legacy to the Eco-sphere community and foundational to my practice and writing.

I. Are We Listening?

The backgrounding of listening (receiving information) is also part of the structure of hyperseparated dualisms …: to speak is the human prerogative (because we have language), it is the active mode of being; listening (or being spoken to) is the passive or recipient position. The power relation is clearly hierarchical: those who speak are more powerful than those who are spoken to. I am proposing that listening, and more broadly, paying attention, should also be considered an active verb; indeed, in multispecies creature communities, it must be so considered. To pay attention is to exercise intelligence, to know so as to be able to inter-act.[19]

Madan-Soni, Roma. *Are We Listening?* (2020).
https://romamadansoni.com/art/;
https://mayinart.com/artists/dr-roma-madan-soni.

---

[19] Rose, 2013, 93-109.

Are We Listening? (2020) demonstrates how considerate listening skills between community members facilitate a just environment. As a queer-antiracist feminist-activist, she aligns her fluid blood-coral-corpus alongside the blood-coral-womb of the ocean. She employs 'peaceful-blue' feminist communication and biological connections to recommend male feminist associates to listen to women, white comrades to listen to people of colour, and queer and straight confederates to listen to transgender campaigners to eliminate human gender inequity. She echoes Donovan's voice, "We should not kill, eat, torture, and exploit animals because they do not want to be so treated, and we know that. If we listen, we can hear them."[20]

Are We Listening? employs feminist hearing-aids to question the interwoven authority of supremacy, dissention, and hierarchy in terms that are concurrently ecological, social, economic, and political. Her diverse locale demonstrates how this learning and listening can influence more effective coalitions for climate righteousness, echoing visions and promises that are concurrently anti-colonial, feminist, queer, and multi-species. The title of my work refers to feminist communiqué intellectuals who have not only looked at whose speech merits attention, but also at who heeds. Speaking connects with authority, information, and supremacy, while listening links with subservience. Feminist approaches highlight listening as an assurance of good learning-listening to one's research subjects, the subjugated, and one's innovative and intellectual community. Also, towards creating assemblies for teamwork, whereby the research subjects can themselves set the schema, direct essentials, and profit from the scholarly enterprise.[21]

Are We Listening? in her human-animal form, like water-Goddesses Yemoja and Ganga, disrupts the disguised, persistent silence around race, colour, and nationality as she immerses herself in the voices/sounds of her community to listen to their attentive presence. She assumes the role of a partaker of indigenous culture who participates in trans-species communications and affinity, shares descent with earth-others: species, and ecologies, plants, wind, water, skies, rocks, and mountains.[22] In contrast, in the Euro-Western culture, Cartesian rationalism overhauls Aristotelianism and Empiricism, whose emphasis is on the fluidity of previous and present-day sensory experiences as sources of world knowledge and understanding. Young children are taught not to listen to or trust communication/information from diverse human communities or the animate world with whom our lives

---

[20] Donovan, 1990, 350-375.
[21] Gaard, 2010, 103-129.
[22] Kimmerer, 2013

intertwine. *Are We Listening?* aligns herself with earth-others: ocean coral, to listen to its bourgeoning and deterioration, share her practices and experiences, and mourn injuries and passing of kin. Their material kinship is grounded in biological science; their communication cannot be validated by modern-day rationalism and technology. Her fluidity can be viewed as archaic 'native primitivism' or perhaps a representation of contemporary everyday existence in fear of rising sea levels.

> It keeps us alert, and alive, and aware, and curious, and engaged at levels... It's the world of life, where the rest of us are engaged in a world of death." ... "There's life here. Respect it." "There's no more sacred sound than the voice of life around me," he told me. "It's the voice of the divine."[23]

A 16,000-year-old pond was ignited by a team of game wardens, killing a female beaver and her offspring, leaving the father behind. Soundscape-ecologist Krause listened to the saddest sound (recorded by his friend) he had ever heard, "coming from any organism, human or other."[24] The lone male beaver swam in slow circles, crying out inconsolably for his lost mate and offspring. *Are We Listening?* derives inspiration from Molnar's painting series, "New Earth 15" (2019-) in which she employs a variety of mediums as languages of community engagement. For Molnar, 'place' is a "form of living history, encompassing all human and other-than-human stories. Place is a record of, and an ongoing, active participant in, struggles for justice, visibility, and vitality."[25] Singaporean-artist Tan Zi-Xi's 20,000 'pieces-of-refuse-suspended-motionless', *Plastic Ocean* (2016) spectacles a disturbingly immersive memento of permanently replaced oceanic-sounds and earth-others. It negotiates cynicism, playfulness, and comicality to connect the spectators with the problem actively. The large scale of the installation immerses spectators in a discourse based on the consequences of ocean anthropocentrism in particular ways: missing sounds of species going extinct, of once-flourishing non-mechanized fishing communities, now conferring bulk fish exhaustion, of imminent danger of coral acidification, and mounting existence and spread of ocean plastic.[26] Xi engages her audience in stories of life and death, at cliff edge, to provide "presence" and "vitality" to "disappearing others."[27] *Plastic Ocean* contests

---

[23] Albeck-Ripka, 2017.
[24] Krause, 2017.
[25] Molnar, 2021.
[26] Tanzer et al., 2015.
[27] Dooren, 2014, 1-8.

Probyn's assertion regarding humans to fulfil their protein requirements by *Eating the Ocean*.[28] However, Buchanan's article "Must We Eat Fish?"[29] maintains that we need to work towards a global cessation of consuming seafood to allow aquatic species to reconstruct and evade extermination, a destiny that numerous species face today.

Are We listening? celebrates multi-species existence, although she mourns human and non-human extinction, and dead zones, "severe discontinuities"[30] of the Anthropocene. Haraway contends that this makes refugees of us all, and proposes "recuperation"[31] through storytelling, to remain with the distress, and the need to recall and honour the disappearing as much as the existing. She invites listeners to heed earth movements: to be guided not only by our heads and scientific knowledge but also by our emotions and the considerable mindfulness that animates all life.[32] The change we need to see requires strategies to be invigorated by environmental politicking, guided by the cognizance of our interconnectedness with the earth. It is time to listen to its resonances and get connected.

II. Are We Sniffing?

Are We Sniffing? (2021) in her blood-coral form holds back putrid oil-spills, reeking-sludge, metal drums of nuclear-waste, and disseminated-plastic from exterminating blood-coral-wombs of the seas and oceans. She immerses into oceanic depths to protect earth's wounded womb from memories of war, ecological destruction, and human fatality. Her seductive immersion in the likes of the Greek-Goddess Thalassa, Aztec-deity Chalchiuhtlicue, and Australian-aboriginal Enigma frees us from the petro-patriarchal, capitalist, industrialist domain for us to *Sniff* the live insignias of the ocean-world.

---

[28] Probyn, 2016.
[29] Buchanan, 2019, 29.
[30] Cunsolo and Landman, 2017, 25.
[31] Haraway, 2016, 160.
[32] Haraway, 2015.

A house.
A car. Lights at night and heat in the winter.
A refrigerator to keep food fresh and a stove for cooking.
A better education and a good job. Modern health care.
Wireless communications. Technology and innovation.
The freedom to focus one's daily activities on something more than mere subsistence. These
are among the many
benefits of modern energy. . . . So why energy? Because energy is vital in our everyday lives.[33]

Madan-Soni, Roma. *Are We Sniffing?* (2021).
https://romamadansoni.com/art/;
https://mayinart.com/artists/dr-roma-madan-soni.

As global warming gazes back at us, the fossil-fuel extractor giants of oil and gas brand themselves as saviours of people, their energy and power suppliers. Oil corporations are not alone in their commitment to energy. Energy is a complex, interconnected word. Energy nurtures life itself, its creation and procreation, "everything in the universe may be described in terms of energy,"[34] including living systems, and human cultures. Like we sense it, and Dagget describes, "Energy's meaning is extensive: it is provided by coal, oil, wind; it is a scientific entity; a metaphor; an indicator of vigour, tinged with virtue. Energy feels trans-historic and cosmic, but it is also material: it pumps through pipelines,

---

[33] ExxonMobil, 2015.
[34] White, 1943, 335.

sloshes in gas tanks, and spins wind turbines."[35] Most crucially, energy is the foundational element of modern-day technology, and we need to dig deeper to find the soil that nourishes its roots and nurtures its branches without hurting the ecosystem. The violence of a fossil-fuelled life is carefully hidden underground to make it appealing to consumers. Nevertheless, "the industrial, capitalist system built to hide its waste, with its subterranean pipes, shunted refuse, oceanic dumps and accounting fantasies, is breaking down. Death is overwhelming the seawalls and flooding back into the carefully sanitized cores of privilege."[36] The Western conviction in the impenetrable body and of bodies being "tucked up like little hobbits into the safety of our burrow"[37] has become progressively challenging to preserve. Leakages can be dealt with either openness or resentment. *Are We Sniffing?* sniffs up the waft of less-anthropocentric creative lifestyle proposals evolving into those that are better acclimatized to regeneration, death, and decay: 'loving your monsters',[38] 'learning to die in the Anthropocene',[39] living 'amid the capitalist ruins',[40] 'posthumous life',[41] 'trying to perish better',[42] queer futurity and failure, making kin instead of babies,[43] or art as 'grief-work' for 'after the end of the world'.[44]

American street-artist Caledonia Curry's Greek-Goddess *Thalassa* (2014) is sixteen feet tall, crowned with entangled seaweed in her hair, adorned with a horseshoe-crab plastron-sternum and a ribcage bordered with wriggling pipefish. Her free-flowing Goddess-like form disperses nebulous, oceanic-species, and skeletal-fish streamers omnidirectionally, portraying her as guardian of the vast network of rivers, seas, and oceans. Nevertheless, their skeletal remains arouse spectators towards the devastation of the 'Florida Deepwater Horizon Oil Spill', and other oil-discharges. As a water goddess, she melts male desires; she 'lives in water', threatens to flow away and liquefy masculine virility. Her gigantic size resists dictatorial suppression and "opens up the borders of a hitherto unknown human productive potential, setting in motion streams of money,

---

[35] Dagget, 2018, 25-44.
[36] Ibid.
[37] Morton, 2013, 489-500.
[38] Latour, 2011.
[39] Scranton, 2015.
[40] Tsing, 2015.
[41] Weinstein and Colebrook, 2017.
[42] Grove, 2015.
[43] Haraway, 2016.
[44] Morton, 2013.

commodities, and workers."[45] She decomposes the capitalists' powers towards amassing and accretion to counteract the new industrious possibilities from turning into new human ease. LeMenager's 'oozing-injury' imagery of a woman grappling her child, while gawking through the stench of oil-stained windows at the Pacific, in the wake of the 1969 Santa Barbara oil-spill, shows American Petro-culture bursting through the oceanic-frame, and the 'modern world' as "a dialogue between oil and water."[46]

Artists Brill and Abidi's *Spike Field View 2* (2001) and Demaray's *Sticks and Stones: The Nike Missile Cozy Project* (2001) were exhibited at "Hotspots: Radioactivity and the Landscape," Centre for Arts, University of Buffalo, together with 18 other artists' works. Radioactive waste with radiation levels 70 times higher than surrounding areas replaces ordinary gravel in New York's Niagara County driveways and parking lots.[47] The 1986 nuclear disaster at the Japanese-Fukushima site sent an unparalleled amount of radiation into the Pacific. Previously, atomic bomb tests and radioactive waste polluted the sea; the consequences are visible even today. In addition to being used as a training ground for nuclear war, from 1946 to 1993, more than 200,000 tons of highly radioactive waste in metal drums, nuclear submarines and ammunition has been tipped into the world's oceans.[48] These petro-nuclear-cultural assemblies of marine environments construct the ocean as an enormous and resistant reserve-space to be exploited in the name of economic movement and fiscal liquidity, as one apparently impermeable to anthropogenic damage.

Are We Sniffing? links our smartphone batteries with dead cows, yaks and myriads of fish floating down Tibet's Liqi river, Amur river at the China-Russia border, and Salar de Uyuni-Bolivia. The toxic alkali-metal lithium that powers our phones, tablets, laptops and electric cars leaked from the Ganzizhou Rongda Lithium mine and wreaked havoc within the local ecosystem. The intensive land and water mining operations in West and South Africa required to deliver diamonds, ivory, bauxite, oil, timber and other precious stones and minerals to world markets, degrade land, reduce air quality and pollute water sources. The result is an overall loss of biodiversity and significant environmental impact on human health. The living ocean and land have been re-formed into vessels that hold and transfer toxic whiffs.

American ballet-dancer, scuba-diver 'Underwater Woman' Christine Ren's gestural and abstracted submerged bodily movements in *Jellyfish*

---

[45] Theweleit et al., 1987, 264.
[46] Orr, in LeMenager, 2004, 208.
[47] Meier, 2018.

Soup and *Blind Spots* (2016) of variegated oceanic flora and fauna shows how, in displacing oceanic-materiality, we are dislocating our own location, since we live in "complex entanglements with water."[49] She points towards 'Blue Ecologies' as a means of basing these embodied, creative exercises in certain historical and critical dynamics, most notably the dialogues of extinction, ecological unity and material contraventions that gush through the 'Anthropocene-controlled' women and oceans.[50] Her underwater packing-material-filled shopping cart resists the ecological shock of climate 'changed', transforming ballet and diving into a study of consumerism, capitalism, and industrial incursions into women's and oceanic communities. These demonstrate far-reaching consequences towards organic and cultural life on the planet, moving toward an "anthropogenic extermination event,"[51] and a possible sixth mass extinction.

III. Are We Watching?

> We can't live and avoid killing; this is something I think has been underexplored in vegan literature. . . . We harm others (humans and nonhumans) in all aspects of food production. Many are displaced when land is converted for agricultural purposes, including highly endangered animals like orangutans who are coming close to extinction as a result of the destructive practices...[52]

A Madan-Soni, Roma. *Are We Watching?* (2021).
https://romamadansoni.com/art/;
https://mayinart.com/artists/dr-roma-madan-soni.

---

[49] Neimanis, 2014, 558-575.
[50] Jeffery, 2019.
[51] Dooren, 2014, 1-8.
[52] Gruen, 2014, 132-133.

As she elevates Buddha's peace-emanating lotus through blue-green-oceanic-waters and immerses her blood-coral-corpus within her trans-species setting, now in a state of immense danger, she asks: *Are We Watching?* (2021) the lotus that personifies enlightenment, regeneration, and rebirth in diverse cultures. Its roots sustain-thicken through murky waters for the blossom to regenerate, symbolic of "blue forest ecosystems:"[53] mangroves, seagrass meadows, and salt marshes, some of the most treasured and prolific coastal bionetworks on the planet.[54]

Are We Watching? reminds us how philosophers and scholars of Western culture, eco-critics and eco-anthropologists employ designated rationalism tools to speculate their understanding of non-human animals, plants and other earthly entities. However, "the master's tools will never dismantle the master's house,"[55] and therefore we need to "rethink"[56] and re-situate our positions to companion non-Western cultures and earth-others. This necessitates a re-alignment in building strategies of seeing "creativity and agency in the other-than-human world around us."[57] This shift works in association with animal studies' scholars[58] and plant studies' researchers[59] who promoted approaches of sympathetic-*Watching*, based on a different language of embodiment, performance, spoor, and sensorial interactivities. *Are We Watching?* suggests tools from cross-cultural non-theism principles and mindfulness of Buddhist practice that originated in India.

In regarding "consciousness rather than embodiment as the basis of human identity"[60] from Plato and Descartes, Westerners have ignored their contextual positions as conjoined and interactive with the life and death-flow of earth-others who have agency. The development of plant studies elucidates a botanic understanding within indigenous cultures, Buddhism, and critical ecofeminism, combined with a Western viewpoint that brings together social and environmental, climate, and interspecies justice. Ecofeminism's related ethical veganism offers a valuable stratagem for making virtuous eating choices for humans, animals and plants; its circumstantial feature is not a form of moral doctrine, nor is it a worldwide law.

---

[53] Pendleton et al., 2012, 156-158; Himes-Cornell et al., 2018, 36-48.
[54] Wylie et al. 2016, 76-84.
[55] Lorde, 2018.
[56] Plumwood, 2009.
[57] Ibid. 2009.
[58] Donovan, 1998.
[59] Wakeupscreaming.com, 2018.
[60] Plumwood, 2004, 46.

Are We Watching? sources its name from Jenny Kendler's installation *Birds Watching* (2018) with a hundred images of variously sized disembodied birds' eyes set across a 40 feet steel frame - all species endangered by climate change, *Watching* us? The multi-coloured concentric circles simultaneously imitate a flower-bouquet and a fanciful shooting-range. Their collective questioning gaze "reverses the customary human-bird watcher/watched dynamic to incarnate something like an interspecies conscience."[61] Spectators immersed in the space *Watch* its material and ethical-susceptibilities and our co-existence within live and dead zones.[62]

Are We Watching? Japanese-artist Kojima's life-size (32m long) washi-papercut suspended blue whale, *Shiro* (2018), glistening in a deep dark, meditative space, simulating the bottomless ocean. Kojima weaves paper and thread to embed her corpus within the fostered fabric of the animal form she re-creates to experience its "Hidden Beauty and ... the processes of change and augmentation," and the oceanic agency. Interlaced within its materiality is a dialogue of the reception of the grey whales across cultures and communities that correspond to its migrant route. However, even though Japan is 'imagined' as a nation with an extended history of whaling, Holm argues that many coastal populations in Northeast Japan did not participate in active whale-hunting until the expiration of the Meiji period in 1912.[63] Based on the concept of 'coeval moral ecologies'[64], indigenous fishermen believed that whales were an incarnation of the sea-gods who would transport fish near the coast. These communities thrived from the presence of whales. They dreaded the ecological contamination whaling produced at the coast, its flora and fauna, and ultimately its death, as evident in the case of Western Japanese whalers. Kojima's work shows how whalers, politicians, and fisheries' scientists break the chain of coeval interaction as they downplay fisherman's knowledge at the global level and present supposedly impartial scientific data that disrupt the whaling-fishing connection. Giggs examines the ecology and the environmental imagination, animals, landscape, politics, and memory to correlate our thinking of ecological and imaginative interactions between a history of sea monsters, ocean abysses, and whale species in the Anthropocene.[65]

During the Portuguese domination of the East African coast between the 16th and 18th centuries, 30,000 pounds of ivory passed through the port of Sofala annually; greater than the slave trade. In the 19th century, "tusk

---

[61] Bury, 2018.
[62] Wuebben and Trueba, 2019, 65-78.
[63] Holm, 2020, 13.
[64] Coevality.com, 2016.
[65] Giggs, 2018.

lust"[66] referred to the European-explorers and Arab-traders who facilitated intensive exploitation of ivory resources. They "supplied large numbers of religious reliquaries and artistic novelties for Christian Europe,"[67] before which native Africans "never hunted the elephant for its tusks."[68] "Every twenty tusks ha[d] been obtained at the price of a district with all its people, villages, and plantations."[69]

Are We Watching? how "conservation can no longer afford to be marginalized, ... today, we need everyone"[70]. Asher Jay's permanent exhibit at the National Geographic Encounter in New York's Times Square – a large-scale wall-mounted installation *Piece of the Planet*, and an immersive, sound-scaped installation called *Message in a Bottle* - spotlight the illegal ivory trade. In the 2013 grassroots group 'March for Elephants', 1.5 million people *Watched* the internationally crowd-funded initiative to rouse public pressure for amending laws that license the ivory trade. Jay illustrated the blood-ivory story on a gigantic, animated digital billboard in the Big Apple-Times Square hub. At the *Faberge Big Egg Hunt* in New York, her oval ornament helped raise money for anti-poaching efforts in Amboseli. *Are We Watching?* how 'CowParade' of over 5,000 cows painted by about 5,000 artists worldwide, including Karim Rashid, Vivienne Westwood, David Lynch, helped raise about $30 million, and 'The Rhinos Are Coming' by sculptor Ashby and her team was a massive boon towards conservation efforts in Africa against wildlife poaching.

IV. Are We Feeling?

Are We Feeling? (2020) embodies the first-born creature who "was of the same size and kind as a man and woman closely embracing. He caused himself to fall into two pieces, and from him a husband and wife were born."[71] Her fluid embodiment sirens a rooted incandescent, organic-hued form, as she replaces the pervasive Western iconography embedded in Hellenic, Nordic, and Anglo-Saxon representations of pale skin, silken-haired women in art and media. She argues for an interdependent self-identity that connects nationality, colour, class, and race, and multi-species cohabitation. She invokes the primeval-spiritual, terrestrial, and cosmological

---

[66] Scigliano, 2002, 157.
[67] Beachey, 1967; Murphy, 2013, 1-17.
[68] Sikes 1971, 310
[69] Stanley, 1989, 240.
[70] Jay, 2017.
[71] Brihadaranyaka Upanishad, 2020.

powers of women to relook at surviving narratives on universal matriarchy and global health.

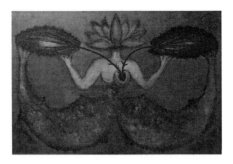

> I closed my eyes and began to feel the rock with my hands—the cracks and crannies, the raised lichen and mosses, the almost imperceptible nubs that might provide a resting place for my fingers and toes when I began to climb. At that moment I was bathed in serenity.... I felt an overwhelming sense of gratitude for what [the rock] offered me—a chance to know myself and the rock differently, to appreciate unforeseen miracles like the tiny flowers growing in the even tinier cracks in the rock's surface, and to come to know a sense of being in relationship with the natural environment. It felt as if the rock and I were silent conversational partners in a longstanding friendship.[72]
>
> Madan-Soni, Roma. *Are We Feeling?* (2020).
> https://romamadansoni.com/art/;
> https://mayinart.com/artists/dr-roma-madan-soni.

There is an evident lack of people of colour, women, and eco-queers as authors of cli-fi narratives. However, feminist, anti-racist, and environmental concepts that corroborate our "transcorporeality with human and more-than-human natures, and our responsibilities/response-abilities to and with these transcorporeal bodies,"[73] persevere. If the realm of global warming narratives in fiction, science-fiction, non-fiction, and film controlled by white men are characterized by "the elision of specific race- and class-based environmental injustices,"[74] then, as Siperstein suggests, we must be "willing to expand our vision of the genre"[75] for artists of colour and eco-

---

[72] Warren, 1990, 125-144.
[73] Gaard's, 2017,
[74] Zizer and Sze, 2007, 387.
[75] Siperstein, 2015.

queers to address global warming. Such action will help us repair/rebuild our relationship with the earth, land, water, and indigenous people, based not on Cartesian-rationalism, but through our *Feelings*.

Are We Feeling? was inspired by Nairobian-American artist Wangechi Mutu's *Water Woman* (2017), who, in her cast bronze ebony skin and large mermaid tail, sits atop a grassy mound gazing East across Laguna Gloria, catching the sun. Rooted in East-African and South American legends of the woman-mermaid Nguva, the "bewitching female aquatic being with powers to entrance and drown susceptible mortals,"[76] and dugong, found in warm coastal waters from East Africa to Australia, the *Water Woman* personifies secrets of a promising Ecocene tomorrow. The cunning temptress, Nguva, deconstructs and disrupts the dominant historical narrative of colonialism and masculinity by assuming a human form to entice, drag colonizers and invaders into the ocean to drown. Her interwoven components of a charcoal-hued form, indigenous legends and Afro-surrealist elements of science-fantasy, critique coloured and female stereotypes and widespread global notions of race, power, and colonialism.

Monarch butterflies, one of the most striking invertebrates, are an emblem of nature: stunning in form, known for their incomprehensible annual migration, and frequent visitors in many backyards. Their conspicuous and swift population declined like a 'silent spring' over the past two decades; reducing biodiversity, part of the present mass extinction, seems un-*Felt*. The acidification of coral reef, the reduction of oceanic species, the polar bear floating on an iceberg in the arctic, or the tear and burn of tropical rain forests, and the expansion of desertification appear un-*Felt*. As legendary Monarch biologist Brower chronologically wrote: In 1986: "Herbicides are a major threat to monarchs, because the butterflies need weeds and wildflowers to survive," In 1990 on deforestation in Mexico, in 2002 on Monarch death triggered by a winter storm in Mexico, in 2006: "The biggest threat to the migration is the steady attrition of forests because of illegal logging,"[77] in 2011 on use of herbicides in agricultural fields, and in 2017 on catastrophic climate change impacts on monarch butterflies.

Are We Feeling? or denying the death of bees based on their articulate-buzzing or missing-humming, as they voice their forced *Savouring* of neonicotinoid pesticides.[78] Or that fish in Athabasca and Slave Rivers and Lake Athabasca voicing ecosystem distress through their tumours, bodily warps, and deformations. In opposition to claims made by industry and

---

[76] Mutu, 2017.
[77] Brower, in Agrawal, 2019, 8093-8095.
[78] Oldroyd, 2007, e168. Termed "colony collapse disorder," a phrase that accuses the bee colonies rather than industrial agrochemicals.

government to the media, the bitumen-oil sands industry significantly increases loadings of toxic PPE to the rivers and their tributaries via air and water pathways.[79] While eco-activists and indigenous peoples convey messages through words, art, and activism, most trans-species communications transmit through images, senses, or *Feelings* rather than words. We can *Feel* their sounds and images and the language of water, wind, and rocks. As Carson, Atwood, Plumwood, Grant, Duvoch, and Shiva, amongst many others, have argued, these value dualisms create value hierarchies that build an artificially definite and higher controller identity by validating the subservience, relegation, and colonization of indigenous people, sexually diverse beings, persons of colour, women, animals, and the natural world.

V. Are We Savouring?

A soil teeming with healthy life in the shape of abundant microflora and microfauna, will bear healthy plants, and these, when consumed by animals and man, will confer health on animals and man. But an infertile soil, that is, one lacking sufficient microbial, fungous, and other life, will pass on some form of deficiency to the plant, and such plants, in turn, will pass on some form of deficiency to animals and man.[80]

Madan-Soni, Roma. *Are We Savouring?* (2020).
https://romamadansoni.com/art/;
https://mayinart.com/artists/dr-roma-madan-soni.

---

[79] Geisy et al., 2019
[80] Howard, in Shiva, 2008.

Are We Savouring? (2020) embraces connected earth bodies as an approach towards planetary justice and survival. To retrieve a critical ecofeminist-interchange, she sources her roots from mythological mermaid-Goddesses, Irish-Sedna and Syrian-Atargatis. She shows that the dialogic intersection of new materialisms, precarious analytical animal and plant readings, and post-humanist studies provides a more comprehensive approach towards communicating contextual and post-colonial feminist methodologies to interspecies justice and culturally situated food/sustenance practices.[81] She points towards real-life enactments and movements, climate justice films, photography, and popular ballads by ecofeminist critics who have unearthed all-encompassing and intersectional narratives. These reinstate condensed chronicles of climate change and offer approaches towards lessening the effects of climate injustices. Mutu's animation film, *The End of Eating Everything* (2013), features singer Santogold as a fantastical, omnipotent Medusa-gorgon who devours the fauna she sees until she self-destroys.

> Soil, not oil, offers a framework for converting the ecological catastrophe and human brutalization we face into an opportunity to reclaim our humanity and our future.[82]

Are We Savouring? (2020) examines Vandana Shiva's proposal, "the food we eat does matter, and makes us who we are, physically, culturally and spiritually."[83] She reminds us how five centuries of colonization and three centuries of fossil fuel extraction distressed the indigenous peoples, lands, waters, and seeds, distorting interspecies relationships. By assigning food sovereignty, sustainability, and seed rights to local farmers around the world and studying their traditional practices and natural healing, we learn that "food and culture are the currency of life,"[84] and that each complements the other. Etched in the chronicles of history, many members of Rajasthan's Bishnoi community lost their lives saving their life-supporting trees after the King's direction to 'fell' them. The 1963 China-India border conflict led to urban development and felling of trees for commercial logging, triggering the widespread floods of the 1970s. The villagers/indigenous communities were connected 'through their umbilical cords' to their oxygen, food, fuel, and shelter, providing trees. Shiva, Gauri Devi, and Bahuguna, with his famous slogan, "ecology is the permanent economy," revived Amrita

---

[81] Gaard 2010, 103-129.
[82] Shiva, 2008, 8.
[83] Shiva, 2021.
[84] Ibid.

Devi's 18th-century ecofeminist 'Chipko Andolan'.[85] "Navdanya"[86] validated how native millets (labelled as "backward" and "primitive" grains by the "the Green Revolution" of the 1960-70s) produced higher yields, better nutrition, and consumed less water than their colonizer-civilized replacements.

In her film, *Wild Relatives* (2018), Syrian artist-filmmaker Manna, from the "musical traditions of Palestine to the global circulation of seeds, the artist unpicks the power structures that support even the most lyrical aspects of life."[87] As Benjamin speculated, "A well-crafted tale plays the long game: 'It resembles the seeds of grain which have lain for centuries in the chambers of the pyramids shut up air-tight and have retained their germinative power to this day."[88] Deep in the earth underneath the Arctic permafrost, seeds from all over the world stored in the Svalbard or Sussex-Wakehurst Global Seed Vaults provide a reserve should catastrophe strike. Manna's film trails the movement and stories of the wild-indigenous seeds and farmers who were displaced by civilized relatives and browbeaten along the way through Syria, Lebanon, to Norway. In 2012 an international agricultural research centre was compulsorily relocated from Aleppo to Lebanon due to the Syrian Uprising-turned war. Planting the seed assemblage from the Svalbard back-ups involved arduous work.

> The sea guides us, and shapes us. It records our behaviours, receives our pollutions. It is fundamental to ceremonies and rituals … [89]

Are We Savouring? Rising Currents: Projects for New York's Waterfront (2010) exhibition at MOMA addressed one of the most urgent challenges facing the nation's biggest city: rising sea levels ensuing from planetary climate change, since "the sea is history-repository of transformation and repair."[90] The Sardar Sarovar Project, India, Hidrosogamoso, Colombia, New Water Culture Movement, Spain, and Lesotho Highlands Project, Lesotho called for social resistance based on diverse social, political, and environmental contexts. Essentially, colonialism commodified and objectified water resources through channelling and colonizing water, as it destroyed community control. This resulted in water deprivation for rural populaces;

---

[85] Embrace Movement, 1973
[86] Millet Festival, 1990. "Gardens of Hope" (2020) sustained Navdanya members during lockdown.
[87] Farzin, 2018
[88] Benjamin, 1936
[89] Walcott, 2007
[90] Helland-Hansen et al., 1995.

in India's case, indigenous water harvesting systems were deliberately demolished.[91] Groundwork and expertise for drinking water facilities were primarily established in areas occupied by the colonizers since European powers were unwilling to invest in their overseas colonies unless it was profitable for them. However, rapidly growing cities such as Bombay in the British Empire experienced deterioration in urban conditions and rising tensions.[92]

Are We Savouring? draws its inspiration from under-ice photographer Lyagushkin's *Two Worlds* (2018) from his White Sea dive. His work shows that even though aquaculture is many thousands of years old, its expansion is essentially a post-World War II phenomenon. In the past 70 years, international aquaculture production has developed nearly 57-fold from 2 to 114.5 million tons in 2018 with a total farmgate sale value of $263.6 billion, accounting for 62.5% of the world's farmed food fish production. More than 300 fish and shellfish species are 'colonized'; the range includes giant clams that acquire most of their nutrients from symbiotic algae, numerous species of carp that are essentially herbivorous, and Atlantic salmon and carnivorous marine fish species.[93] The negative environmental impact that aquaculture has undergone is nuanced, the high density of fish results in the build-up of nutrients and fish waste in the area. Water oxygen gets depleted and leads to algal blooms and dead zones. Antibiotics used by farmers for disease prevention affect the entire ecosystem around the cages and the surrounding wild sea life. The escape of non-native fish causes wild fish to compete for food and potentially displaces native fish.

Conclusions

In the last two decades, climate change has become a leading theme in art and literature and the distinct literary genre of climate change fiction, popularly known as cli-fi. Cli-fi has notably inspired and supported the growth of climate change-oriented studies termed eco-criticism; this includes eco-literature, eco-arts, eco-poetry, and eco-theatres in the media, as well as many large-name productions from 2009 to 2011. The increasing number of eco-critical analyses of climate change literature, especially literary novels, is enhancing climate change fiction. Furthermore, climate change has grown deeper roots within academic concepts, philosophies, and

---

[91] D'Souza, 2006, 621-628.
[92] Nilsson and Nyanchaga, 2009, 105-120.
[93] Williams, 1996

critical theory, as an existentialist problem known as eco-criticism.[94] Eco-criticism reminds us of the animacy of all earth-others, of our historical understanding and symbiotic relationship with animals, plants, soil, water, rocks, and wind - the interruption of autonomous individualism. It reminds us that an animal's cries are not "the striking of a clock"[95] but the agony of another living being who virtues our care and receptiveness; the way that sharing, not covertly amassing resources but using them proportionately according to necessity, generates unity, harmony, balance in the community. "We all do better when we all do better"[96] since we are ecologically, socially, economically and materially co-dependent.

Visual art and textual cli-fi narrative buttress each other to provide new confidence, visualization, and renewed viewpoints that can alter social, political and economic planetary relationships. Artists' artworks remind us that eco-critics' insights are not new. The ideas of our human inter-being with plants, animals, soil, rock, and sky produces relational identities, not autonomous individualism but an animacy of all earth-others. Indigenous trans-species worldviews regarding ecojustice for the future, sourced from the 'elders' have been on the table for human-animals to consume. Indigenous and other communities, whether segregated either ethnically, racially or economically, have a significant contribution to make in ecological, socio-political, and economic terms towards interspecies and climate justice. The value of climate justice accounts in texts, art, audio-visual, and performance is their capability to use art to help people "feel the heat" and activate pro-ecosystem intermediations. *Are We Listening?, Are We Sniffing?, Are We Watching?, Are We Feeling?* and *Are We Savouring?*

References

Agrawal, Anurag A. "Advances in understanding the long-term population decline of monarch butterflies." *Proceedings of the National Academy of Sciences*, 116, no.17 (2019): 8093-8095.

Akhand, Anirban, Abhra Chanda, Kenta Watanabe, Sourav Das, Tatsuki Tokoro, Kunal Chakraborty, Sugata Hazra, and Tomohiro Kuwae. "Low CO 2 evasion rate from the mangrove-surrounding waters of the Sundarbans." *Biogeochemistry* 153, no. 1 (2021): 95-114.

---

[94] Johns-Putra, 2016, 26-42.
[95] Descartes, 1964-75.
[96] Minnesota's progressive Senator Wellstone (1991-2002) coined this phrase.

Albeck-Ripka, Livia. "This is What Extinction Sounds Like." Vice.com, 2017. https://www.vice.com/en/article/paev7v/this-is-what-extinction-sounds-like-v24n5.

Barad, Karen. *Meeting the Universe Halfway: Quantum Physics and the Entanglement of Matter and Meaning*. Chapel Hill, NC: Duke University Press, 2007.

Beachey, Raymond W. "The East African Ivory Trade in the Nineteenth Century." *The Journal of African History,* 8, no.2 (1967): 269-290. http://www.jstor.org/stable/179483.

Berry, Wendall. *Sex, Economy, Freedom & Community*. Pantheon, 1993.

Bloggs, Joe, and Smith, Kate. *A Guide to Referencing*. Newcastle upon Tyne: Cambridge Scholars Publishing, 2019.

Brady, Amy. "Chicago Review of Books - Burning Worlds." *Olympic Climate Action*. Olyclimate.org, 2020. https://olyclimate.org/2020/01/02/chicago-review-of-books-burning-worlds/.

Mark, Joshua J. "Upanishads: Summary & Commentary." *Ancient.eu,* 2020. https://www.ancient.eu/article/1567/upanishads-summary--commentary/.

Buchanan, Ian. "Must We Eat Fish?." *Symplokē* 27, No. 1-2 (2019): 79-90.

Bury, Louis. "The Ghosts of Our Future Climate at Storm King". *Hyperallergic,* 2018. https://hyperallergic.com/451982/indicators-artists-climate-change-exhibition-storm-king-art-center/.

Buchanan, Ian, and Jeffery, Celina. Towards A Blue Humanity. *Symploke,* 27 no.1 (2019): 11-14. https://www.muse.jhu.edu/article/734647.

Climate Change and Nuclear Power 2015. International Atomic Energy Agency. https://www.iaea.org/publications/10928/climate-change-and-nuclear-power-2015.

Coevality.com. "It is time to think about being Coeval." *Coevality.com,* 2016. https://coevality.com/2016/04/13/manifesto.

Cunsolo, A., Landman. K. "Introduction: To Mourn Beyond the Human." *Mourning Nature: Hope at the Heart of Ecological Loss and Grief.* Montreal: McGill-Queen's UP, 2017, 3-25.

Daggett, Cara. "Petro-masculinity: Fossil Fuels and Authoritarian Desire." *Millennium: Journal of International Studies*, 47, no.1 (2018): 25-44.

Descartes, Rene. *Oeuvres de Descartes, ed. Charles Adam and Paul Tannery,* 12 vols. (Paris Cerf, 1897-1910; rpt. Paris: Vrin, 1964-1975), Vol. III, 297-298 (hereafter *Oeuvres*). When no reference is made to a translation, the translation is mine.

Donovan, Josephine. "Animal Rights and Feminist Theory." *Signs* 15 no.2 (1990): 350-375.

D'Souza, Rohan Ignatious. "Water in British India: The Making of a 'Colonial Hydrology'." *History Compass*, 4 no.4 (2006): 621-628.

Exxon Mobile. "The Outlook for Energy: A View to 2040." *Supplychain 247*, 2015. http://www.supplychain247.com/paper/the_outlook_for_energy_a_view_to_2040.

Farzin, Media. "Jumana Manna: The Violence of Beautiful Things." *Frieze*, 2018. https://www.frieze.com/article/jumana-manna-violence-beautiful-things.

Gaard, Greta. *Critical Ecofeminism (Ecocritical Theory and Practice)*. Lexington Books, 2017.

Gaard, Greta. "Speaking of Animal Bodies." *Hypatia* 27 no.3 (2012): 29-35.

Gaard, Greta. "Reproductive Technology, or Reproductive Justice? An Ecofeminist, Environmental Justice Perspective on the Rhetoric of Choice." *Ethics & the Environment*, 15 no. 2(2010): 103-129.

Tendler, Brett, Ehimai Ohiozebau, Garry Codling, John P. Giesy, and Paul D. Jones. "Concentrations of Metals in Fishes from the Athabasca and Slave Rivers of Northern Canada." *Environmental Toxicology and Chemistry* 39, no. 11 (2020): 2180-2195. https://setac.onlinelibrary.wiley.com/doi/10.1002/etc.4852.

Giggs, Rebecca. "Whales and Ocean Voids-Wavescapes in the Anthropocene." *International conference on ecocriticism and environmental humanities*. University of Split, 2018.

Grove, Jairus. "Of an Apocalyptic Tone Recently Adopted in Everything: The Anthropocene or Peak Humanity?" *Theory & Event* 18 no.3 (2015). https://www.muse.jhu.edu/article/586148.

Gruen, Lori. "Facing Death and Practicing Grief," 127-141 in Carol J. Adams and Lori Gruen, eds., *Ecofeminism: Feminist Intersections with Other Animals and the Earth*. New York: Bloomsbury, 2014.

Haraway, Donna J. *Staying with the Trouble: Making Kin in the Chthulucene*. Durham: Duke University Press, 2016, 160.

Haraway, Donna. "Anthropocene, Capitalocene, Plantationocene, Chthulucene: Making Kin." *Environmental Humanities* 6 (2015): 159-165.

Helland-Hansen, E., T. Holtedahl and K.A. Liye, *Environmental Effects* (3) Hydropower Development. Norwegian Institute of Technology, 1995.

Himes-Cornell, A., Pendleton, L., and Atiyah, P. "Valuing ecosystem services from blue forests: a systematic review of the valuation of salt marshes, seagrass beds, and mangrove forests." *Ecosystem Services* 30 (2018): 36-48.

Holm, Fynn. *Living with the Gods of the Sea: Anti-Whaling Movements in Northeast Japan, 1600-1912*. University of Zurich. Dissertation, 2020, 13.

Jay, Asher. "Asher Jay the Wild Creative - National Geographic Explorer." *Artist. Designer. Writer. Speaker. Asherjay.com,* 2019. http://www.asherjay.com/about-asher-jay.

Jeffery, Celina. "Artistic Immersion: Towards an Oceanic Connectedness." *Symploke* 27, no. 1-2 (2019): 35-46. https://www.muse.jhu.edu/article/734649.

Johns-Putra, Adeline. "The Rest Is Silence: Postmodern and Postcolonial Possibilities in Climate Change Fiction." *Studies in the Novel*, 50 no.1 (2018): 26-42.

Kimmerer, Robin Wall. *Braiding Sweetgrass: Indigenous Wisdom, Scientific Knowledge, and the Teachings of Plants.* Minneapolis, MN: Milkweed Press, 2013.

Krause, Bernie. "The Voice of the Natural World." *TED.com,* 2017. https://youtu.be/uTbA-mxo858.

Leikam, Susanne and Leyda, Julia. "Cli-Fi in American Studies: A Research Bibliography." *American Studies Journal*, 62 (2017): 08. http://www.asjournal.org/62-2017/cli-fi-american-studies-research-bibliography/.

LeMenager, Stephanie. "Living Oil." Sheena Wilson, Adam Carlson, and Imre Szeman, eds., *Petrocultures: Oil, Politics, Culture*. Montreal: McGill-Queen's University Press 2017, 208.

Latour, Bruno. "Love Your Monsters: Why We Must Care for Our Technologies As We Do Our Children." *Breakthrough Institute,* 2012. https://thebreakthrough.org/journal/issue-2/love-your-monsters.

Lorde, Audre. "The Master's Tools Will Never Dismantle the Master's House." *Sister Outsider: Essays and Speeches*. Berkeley, CA: Crossing Press, 1984: 110-114.

Mcleod, Elizabeth, Gail L. Chmura, Steven Bouillon, Rodney Salm, Mats Björk, Carlos M. Duarte, Catherine E. Lovelock, William H. Schlesinger, and Brian R. Silliman. "A blueprint for blue carbon: toward an improved understanding of the role of vegetated coastal habitats in sequestering CO_2." *Frontiers in Ecology and the Environment* 9, no. 10 (2011): 552-560.

Meier, Allison. "Artists Confront the Radioactive Landscapes of the United States." *Hyperallergic,* 2018. https://hyperallergic.com/462890/artists-confront-the-radioactive-landscapes-of-the-united-states/.

Molnar, Daniela Naomi. "Artist's Statement." *Danielamolnar.com*, 2021. http://www.danielamolnar.com/about/.

Morton, Timothy. "Victorian Hyperobjects," *Nineteenth-Century Contexts*, 36 no. 5 (2014): 489-500.

Murphy, Ryan Francis. "Exterminating the Elephant in 'Heart of Darkness'." *The Conradian*, 38 no. 2 (2013): 1-17. http://www.jstor.org/stable/24614055.

Victoria Miro. "Wangechi Mutu at the Contemporary Austin." Victoria Miro (2017). https://www.victoria-miro.com/news/777.

Mutu, Wangechi, and Santigold. "The End of eating Everything." Nasher Museum at Duke (2013). https://www.youtube.com/watch?v=wMZSCfqOxVs.

Neimanis, Astrida. "Alongside the Right to Water: A Posthumanist Feminist Imaginary." *Journal of Human Rights and the Environment*, 5 no. 1 (2014): 5-24.

Neimanis, Astrida and Rachel Loewen Walker. "Weathering: Climate Change and the 'Thick Time' of Transcorporeality." *Hypatia* 29 no. 3 (2014): 558-575.

Nilsson, David, and Nyanchaga, Ezekiel Nyangeri Nyanchaga. "East African Water Regimes: The Case of Kenya." *The Evolution of the Law and Politics of Water*, pp. 105-120. Springer Nature Switzerland, 2009.

Nixon, Robert. *Slow Violence and the Environmentalism of the Poor*. Cambridge: Harvard UP, 2011.

Oldroyd Benjamin P. "What's killing American honey bees?" *PLoS biology* 5 no. 6 (2007): e168.

Pendleton, Linwood H., Olivier Thébaud, Rémi C. Mongruel, and Harold Levrel. "Has the value of global marine and coastal ecosystem services changed?." *Marine Policy* 64 (2016): 156-158.

Plumwood, Val. "Nature in the Active Voice." *Australian Humanities Review*, 46 (2009). http://australianhumanitiesreview.org/2009/05/01/nature-in-the-active-voice/.

Plumwood, Val. "Gender, Eco-Feminism and the Environment." Edited Robert White, *Controversies* in *Environmental Sociology*. Cambridge University Press, 2004, 43-60.

Plumwood, Val. "The Concept of a Cultural Landscape: Nature, Culture, and Agency in the Land." *Ethics and the Environment* 11 no. 2 (2006): 115-150. http://www.jstor.org/stable/40339126.

Probyn, Elspeth. *Eating the Ocean*. Durham: Duke UP, 2016.

Rose, Deborah Bird. "Val Plumwood's Philosophical Animism: Attentive Interactions in the Sentient World," *Environmental Humanities* 3 no. 1 (2013): 93-109.

Salmon, Enrique. "Sharing Breath: Some Links Between Land, Plants, and People." Edited by Alison H. Deming and Lauret E. Savoy. *The Colors of Nature: Culture, Identity, and the Natural World.* Minneapolis, MN: Milkweed Publications, 2011, 196-210.

Scencosme-Phonopholium. *Phonopholium* (2018) (Sculpture) Wakeupscreaming.com, 2018. https://wakeupscreaming.com/gregory-lasserre-anais-met-den-ancxt-scenocosme/

Scigliano, Eric. *Love, War, and Circuses.* Boston: Houghton Mifflin Company, 2002.

Scranton, Roy. *Learning to Die in the Anthropocene: Reflections on the End of a Civilization San Francisco.* City Lights Publishers, 2015.

Sekula. Allan, Noel Burch, and Ricardo Tessi. *The Forgotten Space.* Documentary. Icarus Films, 2010.

Siperstein, Stephen. "Climate Change Fiction: Radical Hope from an Emerging Genre." Eco-fiction, 2015. http://eco-fiction.com/climate-changefiction-radical-hope-from-an-emerging-genre/.

Shiva, Vandana. "Vandana Shiva on Why the Food we eat Matters." *BBC.com,* 2021. http://www.bbc.com/travel/story/20210127-vandana-shiva-on-why-the-food-we-eat-matters.

Shiva, Vandana. *Soil Not Oil: Climate Change, Peak Oil and Food Insecurity.* London and NYC, Zed Books. 2008, 8.

Sikes, Sylvia. *The Natural History of the African Elephant.* American Elsevier Publishing Co, New York, 1971, xxv + 397.

Stanley, Henry. *In Darkest Africa.* New York: Charles Scribner's Sons, 1 (1989): 240.

Stone, Erin, and Hornak, Lisa. "Losing Ground." *The Atlantic Selects,* 2019. https://www.theatlantic.com/video/index/591832/climate-refugees/.

Szabo, Ellena Briano. *Saving the World One Word at a Time.* CreateSpace Independent Publishing Platform, 2015.

Theweleit, Klaus. *Male Fantasies, Vol. 1: Women, Floods, Bodies, History,* translated by Chris Turner, Stephen Conway, and Erica Carter, Minneapolis: University of Minnesota Press, 1987, 264.

Tsing, Anna Lowenhaupt. *The Mushroom at the End of the World: On the Possibility of Life in Capitalist Ruins.* Princeton University Press, 2015.

Van Dooren, Thom. *Flight Ways: Life and Loss at the Edge of Extinction.* New York: Columbia University Press, 2014, 1-8.

Walcott, Derek. *Selected Poems by Derek Walcott (2007-12-26).* Farrar, Straus and Giroux, 2007.

Warren, Karen. "The Power and the Promise of Ecological Feminism." *Environmental Ethics*, 12 (1990): 125-144.

Weinstein, Jami and Colebrook, Clair. *Posthumous Life Theorizing Beyond the Posthuman.* Columbia University Press, 2017.

White, Leslie A. "Energy and the Evolution of Culture." *American Anthropologist*, 45 no. 3 (1943): 335.

Wylie, Lindsay, Ariana E. Sutton-Grier, and Amber Moore. "Keys to successful blue carbon projects: lessons learned from global case studies." *Marine Policy* 65 (2016): 76-84.

Wuebben, Daniel, and Juan Jose González-Trueba. "Surfing Between Blue Humanities and Blue Economies in Cantabria, Spain." *Symplokē*, 27 no. 1-2 (2019): 65-78.

Ziser, Michael and Julie Sze. "Climate Change, Environmental Aesthetics, and Global Environmental Justice Cultural Studies." *Discourse* 29 no. 2 (2007): 384-410.

CONTRIBUTORS

Kübra Baysal has a PhD, MA, and BA in English literature and works as a lecturer at Ankara Yildirim Beyazit University School of Foreign Languages. As a graduate of Hacettepe University English Language and Literature Department (2008), she received her MA from Atatürk University from the same department in 2013. She earned her PhD from Hacettepe University English Language and Literature Department in January 2019. Her main fields of interests are climate fiction, apocalypse fiction, Doris Lessing, feminism, environmental studies, the Victorian novel, and the contemporary novel. She has had translation works, book chapters, and research articles published in international books and journals.

Onur Ekler is a lecturer in the English department at Mustafa Kemal University. He earned his Ph.D. from Erciyes University. His work focuses specifically on the modernist/postmodernist Self-studies in the literary canon.

Adrian Tait, Ph.D., is an independent scholar and ecocritic with a particular interest in the Victorian literary response to environmental crisis. He has published related papers in a number of scholarly journals, including *Green Letters: Studies in Ecocriticism*, the *European Journal of English Studies*, and *Cahiers victoriens et édouardiens*, and contributed to essay collections such as *Nineteenth-Century Transatlantic Literary Ecologies* (2017), *Victorian Ecocriticism: The Politics of Place and Early Environmental Justice* (2017), *Perspectives on Ecocriticism* (2019), *Literature and Meat Since 1900* (2019), and *Gendered Ecologies: New Materialist Interpretations of Women Writers in the Long Nineteenth Century* (2020).

Seher Özsert was born in Kayseri, 1988. She is a teacher and writer. After completing her MA in English Language and Literature, she received a PhD degree in the same field. She has also attended numerous educational programs, seminars, and conferences both in Turkey and abroad. She worked in several professional environments as a language and literature instructor. She currently works as an Assist. Prof. of English Literature at Nişantaşı University in Istanbul. She is a passionate literature reader and critique of various types of literary texts in World Literature. Her expertise areas are eco-criticism, science-fiction, feminism, postmodernism and postcolonialism.

Pınar Süt Güngör is currently teaching at Muş Alparslan University, Turkey. She holds a BA from the Department of English Language and Literature of Karadeniz Technical University, Turkey. She received her MA in 2015, and PhD in 2020, both from Atatürk University, Turkey. Her doctorate thesis focuses on African-American author Toni Morrison's four novels (*Song of Solomon, Beloved, Love, A Mercy*) in terms of time-memory theories of French philosopher Henri Bergson. Her main research fields of interest are Modern African-American Novel and Comparative Literary Studies. She has delivered a number of conference papers and journal articles on these issues.

Elvan Karaman is an assistant professor in English Language and Literature Department at Istanbul Ayvansaray University in Istanbul, Turkey. She completed her Ph.D., Master's, and undergraduate studies in the department of English Language and Literature at Ege University in Izmir, Turkey. Her research interests lie in the post-war drama, Elizabethan drama, and postmodern literature.

Niğmet Çetiner graduated from Hacettepe University's Department of English Language and Literature in 2012. She received her MA from Atatürk University's Department of English Language and Literature in 2018. She is currently doing a PhD at Atılım University's Department of English Culture and Literature. She works as a lecturer at Kastamonu University's School of Foreign Languages. She has presented several papers at international conferences and published articles and a chapter on environmental humanities studies, contemporary British fiction, Morgan Llywelyn, and the theories of New Materialisms and the Anthropocene.

Emily Arvay completed her PhD at the University of Victoria in 2019 with her thesis "Climate Change, the Ruined Island, and British Metamodernism." Since then, she has worked as a Learning Strategist and EAL Specialist at the University of Victoria. She is currently conducting further research on the intersections between literary metamodernism and contemporary climate fictions. Her chapter on "Society Islands and the Anthropocene in David Mitchell's *Cloud Atlas* was supported by the Social Sciences and Humanities Research Council of Canada and Ian H. Stewart Graduate Fellowship.

Risha Baruah is a PhD Research Scholar from Cotton University, Department of English. She had been engaged as a former part-time faculty in the Department of English at Cotton University and the Department of English at B. Barooah College. She was also previously engaged as an Academic Counsellor to the Department of English at Indira Gandhi National Open University. Her areas of interests are Ecocriticism, Anthropocene studies, Indian Literature, Animal Studies and Posthumanism. She has authored several chapters for national and international books as well as scholarly articles for national journals. She has written academic content for Krishna Kanta Handique State Open University and Institute of Distance and Open Learning. At present, she is an Editor-cum-Reviewer in the Editorial Board of *Aesthetique International Literary Journal* (AJILE).

Anastasia Logotheti, PhD, is Professor of English at Deree College, the American College of Greece (https://www.acg.edu/faculty/anastasia-logotheti/). She has published several articles in *The Literary Encyclopedia* on Kazuo Ishiguro, Graham Swift, and Ian McEwan. Her most recent publications are the articles "Alterity in E M Forster's 'The Other Boat'" in *Language and Literary Studies of Warsaw* (2021) and "Digital Encounters with Shakespeare" in *Research in Drama Education* (2020) as well as chapters in the volumes *Crossing Borders in Gender and Culture* (2018), *Reading Graham Swift* (2019), and *London: Myths, Tales and Urban Legends* (2021).

Işıl Şahin Gülter is an Assistant Professor of English Language and Literature at Fırat University. She received her PhD from Istanbul Yeni Yüzyıl University, Turkey, in 2018. She has written widely on contemporary British drama. Her recent research interests include climate change theatre, ecodramaturgy, ecocriticism, and ecofeminism. Her most recent publications include work on contemporary authors and playwrights, including Liz Jensen, Sarah Hall, Caryl Churchill and Mike Bartlett. She is currently working on a monograph concerning climate change in contemporary British drama.

Sukanya B. Senapati is a Teaching Professor of English at the University of South Florida, Sarasota Manatee, Florida, USA. Her research interests are in the areas of, World Literature, Ethnic-American Literature, and Environmental Literature. Her most recent works are, "Making Black Lives and Families Matter: Honoring Family and Fatherhood in *God Help the Child.*" Critical Responses About the Black family in Toni Morrison's God Help the Child, ed. by Rhone Fraser and Natalie King-Pedroso, 2020 and "Ramayana's Hanuman – Animal, Human or Divine" *The Human- Animal*

Boundary: Exploring the Line in Philosophy and Fiction, ed. by N. Batra and M. Wenning, 2019, both published by Lexington Books.

Andrew Erickson is doctoral researcher and instructor in the Department of English and American Studies at the Europa-Universität Flensburg and Editorial Assistant for *Amerikastudien / American Studies: A Quarterly*. His current dissertation project understands speculative fiction by Black and Indigenous makers through the lens of the postapocalypses of American enslavement and settler colonialism. Research interests include postcolonialism, science fiction, critical posthumanism, and digital humanities.

Murat Kabak is a research and teaching assistant at the Department of English Language and Literature, Istanbul Kültür University. He received his M.A. degree in English Literature at Boğaziçi University. He is currently a doctoral student in the English Literature program at the Institute for Graduate Studies in Social Sciences, Boğaziçi University. His research interests include contemporary fiction, critical theory and criticism, and film studies.

Manasvini Rai is a Ph.D. research scholar in English, in the Department of Humanities and Social Sciences, at the Malaviya National Institute of Technology, Jaipur, India. Her research interests include subaltern studies, intersectionality, and women's writing, with a specific focus on contemporary black British literature. She has presented papers in these areas at international conferences conducted by the University of Maryland, Idaho State University, and California State University (Long Beach and Fullerton) earlier this year. Rai's papers have been published in the *Journal of Literature and Aesthetics* and *Journal of the Faculty of Arts* of the Aligarh Muslim University.

Roma Madan-Soni is an interdisciplinary eco-feminist artist, art historian and researcher (PhD University of Wolverhampton). She has published articles for journals: Ecofeminism and Climate-Change, IFJP, Craft Research, Art & The Public Sphere, NECSUS, JVAP, and Swasti, among others. She has made presentations at CAA, University of Wolverhampton, AUK, KU, GUST, AOU, UN Habitat, JNU, LSC, Raza Foundation, Beit-Sadu amid others. She has received research-grants/commissions and awards from TAPRI, Routledge, TSCK, KFAS, KISR, UN Habibtat, and AOU. She will present her work at Venice International Art Fair and Florence Biennale 2021. Her Ecofeminist-art has been exhibited at TAPRI, TFAP-Rutgers, Ecoproject.Org, TSCK, DAI-Kuwait, MOMA_Kuwait, and

Masaha-13, to name a few. She chairs the 'Transformative-Education Think-Tank' for CIC-Env.earth- Konrad-Adenauer-Stiftung collaboration.